Applications for Sustainable Energy and Environment 2018

AA008938

Topical at the 8th World Congress on Particle Technology

Orlando, Florida, USA
22 - 26 April 2018

ISBN: 978-1-5108-6980-6

Printed from e-media with permission by:

Curran Associates, Inc.
57 Morehouse Lane
Red Hook, NY 12571

CURRAN ASSOCIATES INC.
proceedings.com

Some format issues inherent in the e-media version may also appear in this print version.

Copyright© (2018) by AIChE
All rights reserved.

Printed by Curran Associates, Inc. (2018)

For permission requests, please contact AIChE
at the address below.

AIChE
120 Wall Street, FL 23
New York, NY 10005-4020

Phone: (800) 242-4363
Fax: (203) 775-5177

www.aiche.org

Additional copies of this publication are available from:

Curran Associates, Inc.
57 Morehouse Lane
Red Hook, NY 12571 USA
Phone: 845-758-0400
Fax: 845-758-2633
Email: curran@proceedings.com
Web: www.proceedings.com

TABLE OF CONTENTS

Some abstracts in this book refer to images that were not made available by the publisher

508290 Fluidized Bed Torrefaction of Biomass Pellets: Process Performance and Product Quality

Abstract:

Biomass is considered the renewable energy source with the highest potential for replacing fossil fuels in the short and medium term. Biomass is available almost everywhere on earth in the form of both renewable raw materials and biogenic wastes generated by human activities, e.g., agricultural and livestock activities, food processing and timber industry. Biomass is the only renewable energy, which is based on sustainable carbon. Therefore, while other renewable resources produce heat and power only, biomass has the further advantage of its possible conversion into chemicals, commodities and liquid fuels for transportation [1, 2], thus competing with fossil fuels in all their applications. Unlike other renewable energy resources, such as solar and wind energy, which are intermittent and, therefore, can meet only a portion of primary energy demand, biomass can ensure continuous power and heat generation. Furthermore, it can also be easily transported and stored to accommodate changes in the output demand; it can be used to generate electric power with the same equipment or power plants that are now burning fossil fuels.

On the other hand, there are currently a number of challenges related to the quality of biomass resources that still prevent them from being used on a large scale for heat and power production. Biomass has, in fact, a lower energy content compared with fossil fuels, which means that a higher load of feedstock is required in the case of biomass-fed plant in order get the same amount of energy when compared to fossil fuels. Moreover, biomass is a relatively bulky material. Typically, biomass bulk density ranges from 80-100 kg/m^3 for agricultural straws and grasses to 150-200 kg/m^3 for woody resources like wood chips and sawdust [3], while conversely the bulk density of coal is about 700 kg/m^3 [4]. As a consequence of this, the volume of feedstock to be handled increases enormously, when biomass is used as a fuel instead of coal, with all the consequent problems that this can bring for logistics, in particular storage and transportation, which are factors that greatly affect profit margins and thus the convenience of a biomass-fed plant. This is the reason why, commonly, it is only economically feasible to transport unprocessed biomass over a distance less than about 200 km [4]. In addition, biomass cannot be stored outdoor without careful protection since it is prone to natural decomposition over time and breakdown with exposure to moisture and pests (e.g., flies and mosquitos), with consequent loss of quality and off-gas emissions; the high moisture content of some kinds of biomass also accelerate the decomposition process. It is worth noting that drying biomass has little benefit for the improvement of biomass storage behavior. In fact, because of its hydrophilic nature, biomass can re-absorb moisture and start to decompose again.

Recent developments in mechanical densification technologies, including pelletization and briquetting, have substantially improved the economics of moving biomass around the globe [4, 5]. Typically, these technologies increase the biomass energy density (MJ/m^3) through the increase of its bulk density (kg/m^3), but have a very little benefit for the improvement of properties such as the high oxygen content or low calorific values (MJ/kg) and the remarkable hydrophobic behavior. The lower heating value (LHV) of currently marketed wood pellets is about 15-18 MJ/kg, which still limits the mixing ratio that can be used in co-firing with coal typically having a lower heating value (HHV) of about 23-28 MJ/kg [6]. Fuel pellets with a very high bulk energy density in the range from 15 to 18 GJ/m^3 and a lower heating value as high as 20-24 MJ/kg can be obtained when torrefaction, a relatively new thermal pretreatment of biomass, is combined with pellettization [6]. Note that wood pellets, which are known to be a very energy-dense biomass fuel, have an energy density that typically range from 7.5 to 10 GJ/m^3 while that of coal is about 18.4-23.8 GJ/m^3.

Basically, torrefaction is a thermo-chemical treatment where biomass is heated in an inert environment up to a temperature ranging between 200 and 300 °C. It is traditionally characterized by low particle heating rate (typically less than 50 °C/min) and a relatively long reactor residence time that typically ranges from 30 to 120 minutes depending on the specific feedstock, technology and temperature. The benefits accomplished by torrefaction include the possibility to convert any hydrophilic lignocellulosic raw materials into a hydrophobic solid, which can be stored outdoors. The bulk density of the torrefied material, however, is generally lower than that of the raw biomass, making transport and storage economically challenging. Therefore, combining torrefaction and pelletization has great potential to upgrade raw biomass to a standard commodity fuel.

So far, two potential pathways, either an upstream (torrefaction before pelletization) or downstream (torrefaction after pelletization) configuration, have been assessed for retrofitting torrefaction within pellet production facilities. In particular, outcomes from a recent study by Kumar et al. (2017) have pointed out that downstream integration could be the better solution since it avoids many of the operational and safety issues associated with the size reduction of torrefied material and its densification [7]. Moreover, torrefaction of wood pellets could also led to a 3- to 4-fold increase in torrefaction reactor throughput when compared to untreated wood chips. As regards the quality of torrefied pellets, instead, findings from literature are rather inconsistent. According to some research works [8, 9], it is difficult to bind torrefied biomass particles without a binding agent, while conversely "wood pellets" torrefied at different temperatures exhibit good durability and no breakage was observed upon the thermal treatment. This suggests that the bonds formed during pelletization outlast the torrefaction quite well. Other research works [10], instead, report that torrefaction after pelletization has a negative impact on the quality of pellets. In particular, it decreases the mechanical durability and the bulk density of pellets, thus affecting their handling and combustion performances.

In this context, an experiment study on the torrefaction of wood pellets has been performed and the results are presented in this work. In particular, torrefaction tests were performed in a lab-scale fluidized bed reactor consisting of a tubular glass column (100 mm inner diameter and 75 mm height) equipped with a 4-mm thick sintered-glass gas distributor. In detail, experimental batch runs were performed at three different temperatures (i.e., 200, 250 and 280 °C) and holding times (i.e., 5, 15 and 30 min) in order to investigate the effect of the main process variables (temperature and time) on both the quality of torrefied wood pellets (i.e., elemental composition, bulk and energy densities, sizes and shape, hydrophobic behavior, mechanical durability and fine particles content) and process performances (i.e., mass and energy yields). To the best of the Authors' knowledge, the torrefaction treatment of wood pellets in fluidized bed reactor has not been investigated so far. Therefore, findings of this work can also useful to highlight potential advantages and drawbacks related to use of this technology in this specific application.

Reference

[1] L. Wu, T. Moteki, A.A. Gokhale, .D. W.Flaherty, F. D. Toste. Production of Fuels and Chemicals from Biomass: Condensation Reactions and Beyond. Chem. 1 (2016): 32-58.

[2] L. Tock, M. Gassner, F. Maréchal. Thermochemical production of liquid fuels from biomass: Thermo-economic modeling, process design and process integration analysis. Biomass Bioenerg. 34 (2010): 1838-1854.

[3] J.S. Tumuluru, C. T. Wright, J. R. Hess, K. L. Kenny. A Review on Biomass Densification Technologies for Energy Application. Biofuels Bioprod. Biorefin, 5 (2011):683-707.

[4] S. Clarke, F. Preto. Biomass Densification for Energy Production. Ontario Ministry of Agriculture, Food and Rural Affairs (2011).

[5] A. Uslu, A.P.C. Faaij, P.C.A. Bergman. Pre-treatment technologies, and their effect on international bioenergy supply chain logistics. Techno-economic evaluation of torrefaction, fast pyrolysis and pellettization. Energy 33 (2008): 1206-1223.

[6] J. Koppejan, S. Sokhansanj, S. Melin, S. Madrali. Status overview of torrefaction technologies. In: IEA Bioenergy Task 32 Final Report (2012).

[7] L. Kumar, A. A. Koukoulas, S. Mani, J. Satyavolu. Integrating Torrefaction in the Wood Pellet Industry: A Critical Review. Energy Fuels 31 (2017): 37-54.

[8] B. Ghiasi, L. Kumar, T. Furubayashi, C. J. Lim, X. Bi, C.S. Kim, S. Sokhansanj. Densified biocoal from woodchips: Is it better to do torrefaction before or after densification? Applied Energy 134 (2014): 133–142.

[9] N. Doassans-Carrère, S. Muller, M. Mitzka. REVE: Versatile Continuous Pre/Post-Torrefaction Unit for Pellets Production. In book: World Sustainable Energy Days Next 2014, Chapter: Part II, Publisher: Springer Fachmedien Wiesbaden, Editors: Gerhard Dell, Christiane Egger, pp.163-170.

[10] C. Nobre, M. Gonçalves, B. Mendes, C. Vilarinho, J. Teixeira. Torrefaction effects on composition and quality of biomass wastes pellets. In book: WASTES: Solutions, Treatments and Opportunities, Publisher: CRC Press, Taylor & Francis Group, Editors: Vilarinho C, Castro F, Russo M, pp.171-177.

Copyright © American Institute of Chemical Engineers. All rights reserved.

508893 A Novel Model to Forecast the Yield and Composition of the Pyrolysis Products: Reaction Kinetics and Hydrodynamics Studya Novel Model to Forecast the Yield and Composition of the Pyrolysis Products: Reaction Kinetics and Hydrodynamics Study

Abstract:

A Novel Model to Forecast the Yield and Composition of the Pyrolysis Products: Reaction Kinetics and Hydrodynamics Study

Shuli Shu[a], Sherif Farag[a,b], Francois Bertrand[a,*], and Jamal Chaouki[a,*]

[a]Department of Chemical Engineering, Ecole Polytechnique de Montreal, P.O. Box 6079, Station Centre-ville, Montréal, QC, Canada H3C 3A7

[b]RMTech for Environmental Solutions Inc., 426-239 Deguire, Montreal, Quebec, Canada H4N 1P2

*Corresponding author

E-mail address: francois.bertrand@polymtl.ca (F. Bertrand), jamal.chaouki@polymtl.ca (J. Chaouki)

Biomass is a renewable and sustainable source for energy and chemical industries. Among thermochemical technologies, pyrolysis is the most promising technique to process such complex biomass feedstocks as it is not highly affected by their variable composition and does not suffer from scale-up issues and limitations. The unique advantages of multiphase reactors, such as their excellent heat and mass transfer performance, and the elimination of radial and axial gradients in concentration and temperature, explain why they are of high interest to implement this technology.

A deeper understanding of the hydrodynamics of such reactors requires their operation under stable conditions, which enhances the heat and mass transfer rates and minimizes the energy requirements. Linking the transport physics obtained from hydrodynamics characteristics with reaction kinetics can be done by simulating the pyrolysis products yield accurately. In this regard, this work concerns the modeling of a fluidized bed reactor operating under pyrolysis conditions, while taking into consideration the kinetics of pyrolysis of one of the most abundant by-products from pulping paper industry, lignin. This proposed model is capable of predicting both the yield of the primary pyrolysis products, biochar bio-oil and biogas and the composition of the liquid product.

Copyright © American Institute of Chemical Engineers. All rights reserved.

507859 Investigation of Sludge Conversion during Gasification Process Using Electrical Capacitance Tomography

Abstract:

Waste management and energy consumption are two major challenges throughout the world nowadays. Sewage sludge, for example, is generated from wastewater treatment plants and usually contains a large portion of organic matters that could be further recovered to produce energy. Recently, converting solid waste to energy through gasification technology is attracting more and more focus, as through gasification, toxic gases such as dioxin could be prevented comparing with traditional incineration process [1]. During gasification process, carbon content in the solid waste decreases gradually as it is decomposed and gasified to CO, CO_2, $CH_4$and other gases with smaller quantities. However, gasification is a highly complex process that is difficult to model without very simplifying assumptions, and experiments are limited by the partial information that can be measured. Therefore, the process efficiency can be improved if more theoretical information is recognized. In order to achieve such improvement, the complex fluid dynamics, including reaction rates and reaction zones should be identified. One step toward this target is detecting the profiles of compositions during gasification.

The main goal of this work is therefore to investigate a new method to monitor online the solid waste decomposition along the gasifier, and consequently understand what are the solid and gas compositions at these positions. This identification can also help to recognize what is the local mass transfer rates inside the gasifier from an experimental point of view, which was not achieved before. The potential use of such measurement is not only to measure the local composites, but to be able to produce data sets for quality validations of numerical models that are typically validated by temperature profiles and output composition, rather than full composition analysis throughout the system.

The methodology of the current study is to use Electrical Capacitance Tomography (ECT) technique to correlate the solid waste conversion with solid permittivity. ECT is a well-known method that has been widely used in pneumatic transport to monitor solid volume fraction cross sectional distribution in enclosed pipes or vessels by measuring the variations in the dielectric properties of the materials [2]. Because the particles composition is expected to change through the gasifier, dielectric properties of the material is changing as well, and the measured permittivity should also change accordingly. By correlating this property with carbon conversion, gasification efficiency and syngas production can be monitored easily and fast during industrial operations and controlled experiments with a non-intrusive technique.

[1] Ong Z, Cheng Y, Maneerung T, et al. Co gasification of woody biomass and sewage sludge in a fixed bed downdraft gasifier[J]. AIChE Journal, 2015, 61(8): 2508-2521.

[2] Zhang W, Cheng Y, Wang C, et al. Investigation on hydrodynamics of triple-bed combined circulating fluidized bed using electrostatic sensor and electrical capacitance tomography[J]. Industrial & Engineering Chemistry Research, 2013, 52(32): 11198-11207.

Copyright © American Institute of Chemical Engineers. All rights reserved.

507958 Development of Iron-Based Fischer-Tropsch Reactor Particle Management and Gas-Solid Fluidized Bed Activation Technology

Abstract:

The slurry bed column reactor (SBCR) with gas-liquid-solid three phase has the advantages of small investment, uniform temperature and easy operation. Therefore, most of the iron based low temperature Fischer-Tropsch (LTFT) technologies choose SBCR. But at the same time, SBCR need to be matched with Fischer-Tropsch heavy wax filtration system and online catalyst replacement technology to achieve wax separation from catalyst particles, and online replacement of fine particle or deactivation catalyst to maintain the catalyst performance.

The traditional SBCR slurry filter technology mainly focuses on filter and filtration process, including filter type selection, filtering precision and blowback frequency. The filtration efficiency decline is mainly due to filter channel blocked by the catalyst fine particles during reactor operation, but the filter can't be replaced under the reaction condition, so the only effective way is to reduce fine particles. The traditional way is online replacement with intermittent operation to discharge part of catalyst including fine particles from the reactor. Our research group proposes new catalyst particles management ideas that based on slurry bed fluid mechanics study, the fine particles can be discharged from top of the reactor continuously, so the backmixing fines in the reactor bed and the filter were deceased. This technology can prolong the service life of filter more than 20%. Besides, for the discharged residue slurry out of the reactor, we can apply magnetic separation technology to separate the ferromagnetic iron catalyst form the discharged residue. It can not only save the investment of traditional wax filtration equipment but also increase the recovery rate of wax by more than 10%.

Traditional catalyst online replacement technologies can discharge the fine catalyst particle and some of the deactivated catalyst intermittently, then the new activated catalyst can be added to the reactor. So the fresh inactive catalyst needs to be activated to the active state which has the Fischer-Tropsch activity. Now, for the iron-based LTFT, the catalyst is activated in an independent SBCR reactor system with the same operating temperature and pressure as Fischer-Tropsch reactor, and some other specific activation process conditions. The activation process is intermittent operation mode about 5-7 day per time to fit the Fischer-Tropsch reactor discharge operation. The activated catalyst should be added to Fischer-Tropsch reactor quickly, and cannot be stored for long time.

If using gas-solid fluidized activation method, the active catalyst does not need to suspend as in SBCR reactor, so the offline catalyst activation is possible, active catalyst storage time can be extended, it is also possible to achieve catalyst online replacement continuously. As a result, the catalyst particle management will be more flexible and controllable. But Iron-based LTFT gas-solid fluidized bed activation research is little.

Our group study gas-solid fluidized bed catalyst activation technology. From the hydrodynamic view, the iron-based LTFT catalyst belongs to A-type particle, so the fluidization is easy. We completed the 100 mm diameter gas-solid fluidized bed reactor cold model test to investigates the state of flow characteristics and hydrodynamics character; According to the physical property characteristics of iron-based catalyst (such as particle density, particle size distribution, Angle of repose, etc.), we study the catalyst fluidization velocity, ensure that the superficial gas velocity within the range of 0.1 to 0.3 m/s can fully fluidize the catalyst particle, we also design a new type of gas distributor which can guarantee the fluidization effect and avoid the dead zone.

Fundamentally, from the activation mass transfer view, in gas-solid phase, the resistance is smaller than in three-phase SBCR, so the gas-solid activation may shorten the activation time. The same as SBCR activation, gas atmosphere, temperature, pressure also can affect gas-solid reactor activation result. So we also did the fluidized activation process test in fluidized reactor loading 1.0-5.0 kg catalyst, A variety of activation atmosphere (pure H_2, CO and the mixed H_2 + CO) was carried on the system research, it was found that atmosphere has a great influence on catalyst activity and stability. When the H_2/CO ratio of activation fresh syngas is between 5/1 and 20/1, catalyst will get a comprehensive performance. Test results also show that the catalyst activation temperature and time should match, proper temperature not only ensure the performance of the catalyst, also reduce activation time, the reasonable temperature range is 250-270°C, the corresponding activation time is 6 to 12 hours, while the existing slurry bed activation requires 24h. On the other hand, if the activation temperature exceeds 280°C, catalyst will be sintered to some extent and catalyst performance is poorer. Compared with SBCR activated catalyst, Gas-solid fluidized activated catalyst almost has the same Fischer-Tropsch catalyst performance: the CO conversion and C_5^+ hydrocarbon selectivity difference is within 2%, methane selectivity is within 0.5%, CO_2 selectivity within 1%. At present, the crystal structure transformation of catalyst under different activation process parameters is study, and the precise control of the activity will be achieved finally.

Above all, the new catalyst particles management is the core in NICE LTFT technology, and we have developed a SBCR reactor fine particle treatment process and gas-solid fluidized bed catalyst activation technology. Now we are planning to build a 20 tons/batch fluidized bed activation demo unit.

Keywords: Fischer-Tropsch technologies; Slurry bed column reactor; Gas solids fluidized bed; Flow behaviors

Copyright © American Institute of Chemical Engineers. All rights reserved.

507026 Challenges and State of Art of Multiphase Flow Modeling in F-T Synthesis

Abstract:

Modeling the gas-liquid-solid flow is critical for the scale-up and optimization of slurry bubble column reactors in Fischer-Tropsch (F-T) Synthesis. This is, however, a challenging task for CFD modelers due to inadequate understanding of the complex mesoscale flow structure and the complex coupling of multiphase flow, interfacial mass and heat transfer as well as reaction kinetics. This talk gives a state of art review in literature and the challenging issues of multiphase flow modeling in F-T synthesis. The challenges in physical models involve the phase interaction (drag) and multiphase turbulence modeling, and bubble breakage and coalescence modeling. The challenges in reactor modeling involve the simulation of complex geometry of a large number of internal cooling tubes, the effects of gas contraction due to reaction on fluid dynamics and the models for catalyst particle attrition. A mesoscale modeling approach for modeling multiphase reactors is then developed based on the Energy-Minimization Multiscale (EMMS) methods, and a stability-constrained multi-fluid model is established to provide closure models for CFD modeling.

Copyright © American Institute of Chemical Engineers. All rights reserved.

508043 A Transient DEM-Based Virtual Experimental Blast Furnace Model Realized through Scaling

Abstract:

A transient DEM-based virtual experimental blast furnace model realized through scaling

Qinfu Hou,[1] Dianyu E,[1] Shibo Kuang,[1] and Aibing Yu[1,2]

[1]ARC Research Hub for Computational Particle Technology, Department of Chemical Engineering, Monash University, Clayton, VIC 3800, Australia

[2]Centre for Simulation and Modelling of Particulate Systems, Southeast University - Monash University Joint Research Institute, Suzhou 215123, PR China

Abstract

Intensive heat and mass transfer between continuum fluids and discrete particulate materials is quite common in many chemical processes. To understand and improve the operation of these processes, discrete particle models are very helpful when they are combined with the flow, heat transfer and chemical reaction models. However, it is not practical to predict the transient process due to the high computational cost. Here, a transient discrete element method based virtual blast furnace model is developed through scaling. Such a scaled model can simulate the process two orders faster. The scaled model is first introduced and verified extensively for the prediction of flow, heat and mass transfer with different scaling factors. Then, the model predictions are tested against available experimental results and more sophisticated computational fluid dynamics model predictions under comparable conditions. The results show that the scaled model can well predict in-furnace flow state, temperature distribution, and the characteristics of the cohesive zone. Finally, a discussion of further development is presented. The scaled model can be used to study the effects of burden distribution, inlet gas composition and material properties on the operation and energy efficiency of a BF. It might be modified for other chemical processes such as fluidized beds and rotary kilns not only for better fundamental understanding but also for better process design and control.

Keywords: Blast furnace, Heat and mass transfer, Chemical reaction, Discrete element method, Computational fluid dynamics, Scaling

Copyright © American Institute of Chemical Engineers. All rights reserved.

507874 Study on the Coke Distribution on Catalyst for MTO Fluidized Bed Reactor-Regenerator System

Abstract:

The methanol to olefins (MTO) process provides an alternative approach to produce light olefins from nonoil resources. In industry, the fluidized bed reactor-regenerator system is applied for MTO process, where the SAPO-34 catalyst particles are circulated between the reactor and regenerator. During the circulation, coke deposits on the catalysts in the fluidized bed reactor and is burned off in the fluidized bed regenerator, which lead to that coke content on catalysts demonstrates a certain distribution in the system. For MTO process, the nonlinear relation of reaction rates with coke content on catalysts indicates that the catalytic performance of the reactor is highly dependent upon the coke distribution of catalysts in the reactor.

In our recent work, a coke distribution model of MTO process was developed based on our new lumped kinetic model. The kinetic model for SAPO-34 catalyst, based on the dual-cycle reaction mechanism, was established from our laboratory fluidized bed experiments. In this kinetic model, a deactivation function was proposed to account for the deactivation process of SAPO-34 catalysts. Then, three dimensional (3D) partial differential equations of coke distribution based on population balance model was developed, and analytical coke distribution expressions of steady state were subsequently obtained under the perfectly mixed assumption of the catalysts in the beds. These expressions indicates that the probability density functions of coke distributions of the reactor and regenerator are determined by three factors, i.e. catalyst residence time, coke deposition (or burning) rate and coke distribution of catalyst inflow. The first two factors could be further merged into a dimensionless number, when the rate of coke deposition (or burning) could be expressed in first order relation of coke content.

Compared with age distribution approach, which has been used for MTO process, coke distribution approach shows the physical picture in a clearer way. Most interested properties of MTO process, such as product selections, MeOH conversion and average coke content, could be directly estimated from the profile of coke distribution. Especially, when encountering a wide coke distribution of catalyst inflow, or investigating space-dependent coke distributions of a fluidized bed, the coke distribution approach possesses a distinct advantage.

The simulations based on the coke distribution model of three MTO reactor-regenerator systems, i.e. pilot-scale, demo-scale and commercial-scale with reactor diameters of 0.261, 1.25 and 10.75 m respectively, were performed. The simulated results were in good agreement with the operation data. The effect of coke distribution on MTO reaction rates, as well as the influence on product selections, was also investigated. The use of coke distribution could be helpful in optimizing the operation conditions and the reactor design of the reactor-regenerator system.

Copyright © American Institute of Chemical Engineers. All rights reserved.

508240 Simulation Study on the Reaction-Diffusion Coupling Processes in Simple Pore Structures

Abstract:

Simulation study on the reaction-diffusion coupling processes in simple pore structures

Yanping Li [a,b,c], Mingcan Zhao [c,d], Chengxiang Li [c*], Wei Ge [b,c,d*]

[a] *School of Chemical Engineering and Technology, Tianjin University, Tianjin 300072, China*

[b] *Collaborative Innovation Center of Chemical Science and Engineering (Tianjin),*

Tianjin 300072, China

[c] *State Key Laboratory of Multiphase Complex Systems, Institute of Process Engineering, Chinese Academy of Sciences, Beijing 100190, China*

[d] *School of Chemistry and Chemical Engineering,*

University of the Chinese Academy of Sciences, Beijing 100049, China

E-mail addresses: liyanping@ipe.ac.cn (Yanping Li), licx@ipe.ac.cn (Chengxiang Li), wge@ipe.ac.cn (Wei Ge).

Keywords: molecular simulation, HS-PPM, reaction-diffusion coupling, pore

Nanoporous materials have attractive and wide applications and vital importance in many chemical processes because of the controlled porosity and large surface area. Their performance would be significantly affected by the diffusion and reaction processes insides the pore structures. To achieve high accuracy and high efficiency with computer simulations, a coupling method is introduced to investigate such processes.

Molecular dynamic simulation combing hard-sphere (HS) and pseudo-particle (PP) has been developed [1, 2]. The diffusion in a reaction-free area can be simulated by HS without considering the structure of the reactants. However, the HS model has poor scalability for parallel simulation due to its event-driven algorithm. To improve the scalability of the HS model, PP modeling (PPM), a time-driven modeling approach is coupled with HS model to construct a scalable parallel and high efficiency method through domain decomposition. Therefore, HS-PPM has the potential to be applied to large-scale parallel computing of diffusion-reaction processes inside nanoporous materials with complex structure.

The HS-PPM coupling method is first applied to study the effect of coke deposition on the diffusion of methane in nanoporous zeolite ZSM-5 (MFI) [2]. The reaction-diffusion coupling leads to a complex coke distribution in zeolite, which has a significant influence on gas diffusion process. Two different coke distribution models are proposed to simulate the actual coke formation process. The results show that a better compromise of reaction and diffusion process in pores leads to a uniform distribution of coke deposition, which has little influence on gas diffusion, especially for a low amount of coke. On the contrary, a worse compromise will result in a concentrated distribution of coke at some active sites, which block the diffusion of gas in the pore completely. Therefore, the coke will have significant influence on gas diffusion, even for a little amount of coke deposition.

Then the HS method is applied to study the reaction-diffusion coupling processes in several simple pore structures with the same volume, including the straight, T-shaped and cross-shaped pores, to understand the coupling mechanism. For a first-order isothermal reaction A¡úB, a diffusion factor *D* and reaction factor *R* are proposed to quantitatively characterize the reaction and diffusion performance of the pore structures. In terms of the proposed factors, *D* and *R*, the optimal structures are investigated through optimizing the overall productivity of these structures. The results show that the productivity depends on the competition of *D* and *R*, which are both determined by the size and shape of the pore structures. The results are expected to be useful for understanding the reaction-diffusion coupling mechanism in the pore structures.

These simulation results demonstrate the effectiveness of the simulation approach used for evaluating the performance of the simple pore structures for simple reactions and the potential of its application in more complicated and practical cases. It also suggests the effectiveness of the proposed factors, *D* and *R*, for charactering the diffusion and reaction processes at molecular level.

References

1. Zhang CL, Shen GF, Li CX, Ge W, Li JH. Hard-sphere/ pseudo-particle modeling (HS-PPM) for efficient and scalable molecular simulation of dilute gaseous flow and transport, *Mol Simulat.* 2016;42:1171-1182.

2. Li YP, Zhang CL, Li CX, Liu ZC, Ge W. Simulation of the effect of coke deposition on the diffusion of methane in zeolite ZSM-5, *Chem Eng J.* 2017;320:458-467.

Copyright © American Institute of Chemical Engineers. All rights reserved.

508474 Enhanced CO_2 Conversion to CO By Silica Supported Perovskite Oxides at Low Temperatures

Abstract:

Repurposing of CO
gas shift chemical looping (RWGS-CL) is capable of efficient CO_2 to CO conversion at a low temperature of ~600 °C with unprecedented rates using the $La_{0.75}Sr_{0.25}FeO_3$ (LSF) perovskite-type oxide amalgamated with silica. The LSF:SiO_2 composite (25% LSF by mass) promotes a notable extent of oxygen vacancies in the active phase, a key parameter for CO_2 conversion. In each of eight RWGS-CL cycles, CO generation yields of LSF:SiO_2 surpass those of LSF alone by about 200%, producing 2.60 mmol CO g_{LSF}^{-1} at a peak rate of 0.8 mmol CO g_{LSF}^{-1} min^{-1}. This significant improvement is concomitant with a decreased average LSF crystalline size retained at these low thermochemical reaction temperatures. Evidence of this enhancement points to wettability by silica, perovskite lattice strain induced by the support, and curtailed quantities of secondary phases that limit accessibility to active surfaces. In this contribution, an appropriate stable platform for improving earth abundancy in perovskite-based redox materials is demonstrated for potential industrial-scale, low temperature CO_2 thermochemical conversion.

Copyright © American Institute of Chemical Engineers. All rights reserved.

510214 Internally Circulating Fluidized-Bed Reactor for Inherent CO_2Â Capture Using Chemical Looping Combustion

Abstract:

Chemical looping process is one of the most promising technologies for CO_2 capture on terms of efficiency and economy. Hence, it provides the means to convert fossil fuel to electricity and/or chemical products with inherent separation of CO_2 and without significant energy penalties. Pressurized operation of chemical looping based processes is however necessary for maximizing their energy efficiency and speeding up their industrial deployment.

This study provides the prospects and opportunities that exist for novel application of the internally circulating reactor (ICR) to be used as an alternative to the conventional dual fluidized-bed reactor. The concept consists of a single reactor that combines two sections, (i.e., the fuel and air sections) with internal physical separations. Specifically the concept aimed to simplify the design, ease the solid circulation, avoid using costly gas/solid separation system, (i.e., cyclone and loop seals), operate at high pressure easily in a single pressurized vessel and eventually bring the chemical looping technology a step closer to commercialization.

Hydrodynamic study carried out in a pseudo-2D cold-flow unit have confirmed that the principle of ICR can be successfully applied to a chemical looping technology process, with easy control of solid circulation and minimum gas leakage between the two sections [1,2]. In this study a hydrodynamic study carried out in the 3D ICR reactor under cold conditions to identify an operation window in which solid circulation is stable and gas leakage is minimum. The effects of the static bed height, bed material properties and gas fluidization velocities on solid circulation rate and gas leakage were determined. The solids circulation rate in the 3D reactor is quantified based on the correlation extracted in the pseudo 2D experiments linking solids circulation rate to the pressure difference between the reactor chambers. Ergun equation is used as a verification of the accuracy of the estimated solids circulation rate based on the measured gas leakage and pressure difference between the reactor chambers. The same equations are used to design the operation window for the ICR under CLC conditions. Reactive experiments are ran to validate the Ergun equation prediction of solids circulation rate by using pressure and gas composition measurements.

Finally, scaling laws are used to translate the operating window designed via hydrodynamic experiments to reactive experiments with the aim of maximizing the fuel conversion and CO_2 capture efficiency. Reactive CFD simulations are completed to verify the accuracy of this scaling and some initial reactive experimental results will be presented to quantify the validity of these methods.

References:

1. A. Zaabout, S. Cloete, S. Amini, Innovative Internally Circulating Reactor Concept for Chemical Looping-Based CO2 Capture Processes: Hydrodynamic Investigation, Chemical Engineering & Technology, 39 (2016), pp. 1413-1424.
2. S. Cloete, A. Zaabout, and S. Amini, The Internally Circulating Reactor (ICR) Concept Applied to Pressurized Chemical Looping Processes. Energy Procedia, 2017. 114 (Supplement C): p. 446-457.

Copyright © American Institute of Chemical Engineers. All rights reserved.

508002 Single-Step Conversion of Methane to Liquid Hydrocarbon Fuels with a Layered Composite Catalyst

Abstract:

Combination of steam reforming and Fischer-Tropsch Synthesis (FTS) is a potential process to convert methane directly to liquid hydrocarbon fuels. In this study, the combination of zeolite membrane coated composite steam reforming catalyst and FTS catalyst was used for a single-step conversion of methane to hydrocarbons. The layered reforming catalyst (zeolite membrane coated $NiMg/Ce_{0.6}Zr_{0.4}O_2$) consisted of an outer layer of H-Î² zeolite encapsulating an inner reforming catalyst core. The layered reforming catalyst was synthesized using a double physical coating method. The layered reforming catalyst has been utilized for methane steam reforming and the methane conversion could reach ~ 30%. The methane conversion decreased with the increase of the zeolite shell thickness, which is proportional to the zeolite amount added [1]. The layered composite catalyst and Fe/SBA-15 catalyst were loaded in a micro fixed-bed glass U-tube. Methane and steam were used as reactants. The primary results show a high methane conversion (> 20%) and hydrocarbons were produced. The operation conditions including methane-to-steam molar ratio, temperature, and gas hourly space velocity (GHSV) will be optimized to produce more hydrocarbons.

[1] Cimenler, Ummuhan, Babu Joseph, and John N. Kuhn. "Effect of Zeolite Membrane Shell Thickness on Reactant Selectivity for Hydrocarbon Steam Reforming Using Layered Catalysts." Energy & Fuels 30, no. 7 (2016): 5300-5308.

Copyright © American Institute of Chemical Engineers. All rights reserved.

508487 Earth Abundant Perovskite Oxides for Low Temperature CO_2 Conversion

Abstract:

The current energy and environmental scenario demands acute attention towards generation of energy in a sustainable fashion. Over the last decade, significant efforts have been spent for the reduction of carbon dioxide emissions and its impact on global climate concerns. Carbon dioxide capture and sequestration (CCS) has garnered a lot of interest leading to several CCS plants being commissioned. However, the scale of CCS (~ 35 Mtpa, in early 2017) is still way below the scale of global CO_2 emissions (~ 35 Gtpa, in 2015). The other promising approach to reduce the atmospheric CO_2 content is thus to repurpose waste CO_2 emissions. Amongst the several possible routes being researched, the solar thermochemical cycles (STC) presents one of the highest sun-to-fuel conversion efficiency.[1] The disadvantages of this process are however high temperatures of operation (more than 1000 °C) and stability of the materials. Reverse water gas shift chemical looping (RWGS-CL) is a modified approach of STC that can convert CO
of ~ 500 °C.[2-3] The RWGS-CL process involves two consecutive steps of reduction and oxidation. The first step consists of reducing mixed metal oxides like perovskite oxides (ABO_3) to their oxygen deficient forms ($ABO_{3-\delta}$) by heating in presence of hydrogen, while the second step involves the conversion of CO_2 to CO over these oxygen vacant materials along the with latter getting converted to their original stoichiometric forms. Since these materials are capable of converting CO_2 to CO, a core component of syngas (a mixture of CO, H_2), they present an opportunity to close a synthetic carbon cycle, enabling carbon footprint reduction of several carbon-positive processes.

An intrinsic parameter that describes the CO_2 conversion capability of these perovskite oxides is their oxygen vacancy formation energy (E_{vac}).[3-4] Since this E_{vac} is closely related to the elemental composition of these perovskites, it is quite rational to tune the perovskite composition for predicting materials with desired CO conversion performance. Perovskite oxide type materials (ABO_3) are perfect platforms for material composition tuning as they can accommodate a variety of elements in their respective 'A' and 'B' sites. Density functional theory (DFT) calculated E_{vac} is thus used to screen different perovskites oxides of the form ABO_3, $A1_{0.5}A2_{0.5}BO_3$, $AB1_{0.5}B2_{0.5}O_3$, and $A1_{0.5}A2_{0.5}$ $B1_{0.5}B2_{0.5}O_3$ (A = La, Ca, Sr, Ba and B = 3d elements from Cr to Ni along with Al and Ga). $La_{0.75}Sr_{0.25}FeO_3$
isothermal CO_2 conversion process via RWGS-CL. Hence, any materials with E_{vac} values close to 3.4 eV were hypothesized to have CO_2 conversion capabilities. Based on this principle, we predicted several perovskite oxides. These were synthesized via Pechini method and then subsequently tested for CO_2 conversion. Amongst many of the predicted materials that demonstrated successful conversion of CO_2 to CO, the earth abundant lanthanum and calcium based oxides showed the best performance till date. We could thus have the highest CO_2 conversion rates at lowest temperatures of ~450 °C by RWGS-CL. Several consecutive cyclic experiments revealed long term stability of these materials enabling potential industrial implementation. An empirical model has also been developed correlating the DFT-calculated oxygen vacancy formation energy of a material to its

intrinsic properties like enthalpy of formation and bond-dissociation energies. This model may allow future prediction of E_{vac} of similar perovskite oxides without rigorous DFT calculations. This RWGS-CL process and earth abundant perovskite material combination can pave the way for sustainable production of CO enabling thermal integration with Fischer Tropsch synthesis (FTS) reaction for the generation of high value hydrocarbons.

References:

1. Chueh, W. C.; Falter, C.; Abbott, M.; Scipio, D.; Furler, P.; Haile, S. M.; Steinfeld, A., High-Flux Solar-Driven Thermochemical Dissociation of CO_2 and H_2O Using Nonstoichiometric Ceria. *Science* **2010,** *330*, 1797-1801.

2. Daza, Y. A.; Kent, R. A.; Yung, M. M.; Kuhn, J. N., Carbon dioxide conversion by reverse water gas shift chemical looping on perovskite-type oxides. *Industrial & Engineering Chemistry Research* **2014,** *53*, 5828-5837.

3. Daza, Y. A.; Maiti, D.; Kent, R. A.; Bhethanabotla, V. R.; Kuhn, J. N., Isothermal reverse water gas shift chemical looping on $La_{0.75}Sr_{0.25}Co_{(1-Y)}Fe_YO_3$ perovskite type oxides. *Catalysis Today* **2015,** *258, 2*, 691-698.

4. Maiti, D.; Daza, Y. A.; Yung, M. M.; Kuhn, J. N.; Bhethanabotla, V. R., Oxygen vacancy formation characteristics in the bulk and across different surface terminations of $La_{(1-x)}Sr_xFe_{(1-y)}Co_yO_{(3-d)}$ perovskite oxides for CO_2 conversion. *Journal of Materials Chemistry A* **2016,** *4* (14), 5137-5148.

Copyright © American Institute of Chemical Engineers. All rights reserved.

539154 Hydrogenolysis of 5-Hydroxymethylfurfural to 2,5-Dimethylfuran Over Supported Pt-Co Bimetallic Catalysts Under Mild Conditions

Abstract:

Highly dispersed Pt-Co bimetallic catalysts were deposited on multi-walled carbon nanotubes (MWCNTs) by atomic layer deposition. High-resolution TEM and TPR analysis verified the formation of Pt-Co bimetallic particles. Catalysts were applied for hydrogenolysis of 5-hydroxymethyfurfural (HMF) to 2,5-dimethyfuran (DMF). High yield of DMF (> 90%) was achieved in hydrogenolysis of HMF over the optimized Pt-Co/MWCNTs catalyst after 8 hr of reaction time under 10 bar H_2 at 160 °C. Through a series of experiments and comparison, the synergistic effect among Pt, Co, and MWCNTs was investigated. The results revealed that the synergistic effect between Pt-Co and MWCNTs played an important role in the improvement of selectivity to DMF for Pt-Co/MWCNTs bimetallic catalysts. In addition, steric hindrance appeared when the Co loading in Pt-Co/MWCNTs was high and it affected the activity of Pt-Co bimetallic catalysts. However, the moderate activity can inhibit the production of byprodct and thereby improve the yield of DMF.

Copyright © American Institute of Chemical Engineers. All rights reserved.

524844 Atomic Layer Deposition (ALD) Coatings Enable Higher Energy Batteries with Enhanced Lifetimes

Abstract:

Lithium-ion batteries are driving todayâ€™s growth in the battery market. Dramatic changes and industry expansion are creating large new markets for cost-effective, nano-engineered materials. ALD Nano uses proprietary Atomic Layer Deposition (ALD) techniques to enable new battery materials that solve battery performance challenges cost-effectively at scale.

Copyright © American Institute of Chemical Engineers. All rights reserved.

506273 Increasing Durability of Fuel Cell Catalysts with Atomic Layer Deposition

Abstract:

The proton exchange membrane (PEM) fuel cell is a highly efficient, portable, non-greenhouse gas emitting source of electrical power that holds great promise as a replacement for the internal combustion engine. One of the most significant challenges in the development of the PEM fuel cell is the long-term durability of the oxygen reduction reaction (ORR) catalyst. Cathode potential cyclingâ€"resulting from the variable voltage loads imposed during vehicular operationâ€"is known to promote agglomeration of the Pt nanoparticle catalyst and degradation of the carbon support. In this work, atomic layer deposition (ALD) was used to produce both modified Pt nanoparticle catalysts as well as PtNi extended surface catalysts with enhanced electrochemical durability. Commercial and ALD-synthesized (on functionalized carbon black) Pt/C catalysts were modified with TiO_2 ALD to create protective nanostructures on the support surface that acted as physical barriers to Pt agglomeration and preserved electrochemical surface area and mass activity over voltage cycling. Extended surface catalysts synthesized by galvanic displacement have previously been shown to have high activity and enhanced durability over Pt/C catalysts due to highly-coordinated surface Pt and elimination of the carbon support. An analog of the galvanic displacement process was developed using Pt ALD on Ni nanowires substrates to create a novel extended surface catalyst with mass activity almost 3X higher than benchmark Pt/C and over 70% retention of activity after durability testing. The PtNi nanowire catalysts were characterized at 5, 10, 20, and 30 cycles of ALD to analyze the nucleation and growth of the Pt during the deposition process and its effects on active surface area.

Copyright © American Institute of Chemical Engineers. All rights reserved.

507254 Evaluating the Influence of Dry Granulation Processing Variables on the Rheological Properties of Granules

Abstract:

Granulation is often employed in the production of pharmaceutical formulations for oral solid dose delivery. It is most often required when constituent materials in a formulation have poor flow properties, enabling these components to be combined to improve process performance and ensure content uniformity.

Dry granulation, based around the use of a roller compactor and integral mill/screen, can have significant advantages over more common wet granulation techniques. Dry granulation can reduce processing costs and can also be used with moisture/solvent/heat sensitive materials. Unlike a wet granulate, the dry granulate does not require additional drying and milling before encapsulation or tabletting.

Dry granulation is not suitable for all types of material, however, there is little research into which properties make a formulation suitable for this process. In addition, there is little to indicate which process parameters produce optimal granulate quality in order to achieve interruption free processing and high quality final products.

In this study an FT4 Powder Rheometer® (Freeman Technology, Gloucestershire, UK) was used to demonstrate how processing parameters influence the properties of a placebo formulation granulate. The granulate was produced using a roller compactor (MiniPactor, Gerteis Maschinen+Processengineering AG, Switzerland) on which the roll gap and compaction force were varied. The dynamic flow, bulk and shear properties of the resulting granulates were then evaluated using the FT4 Powder Rheometer.

The data clearly demonstrates that several rheological properties, known to influence die filling efficiency and Critical Quality Attributes (CQAs) of the resulting tablets, could be controlled by changes in roller compactor settings.

This study emphasises the importance of assessing powder properties using techniques that closely simulate process conditions and how it is possible to identify Critical Process Parameters to optimise a dry granulation process and ensure high final product quality.

Copyright © American Institute of Chemical Engineers. All rights reserved.

507489 Assessment of the Homogeneity of an Alumina Powders Mixture Using Rheological Parameters

Abstract:

This study has been carried out in order to improve the comprehension of powder mixing in the nuclear fuel manufacturing process. Thus, a binary mixture of two Alumina (Al_2O_3) powders, GE15 and CR6, was used to simulate the oxides powders actually involved in the process. Different mixing technics and parameters were tried and compared in terms of homogeneity of the mixtures produced. The concept of intensity of segregation [1] was used for assessing mixture homogeneity through a sampling protocol and classical statistical analysis tools [2]. One of the difficulties of working with these compounds is that, due to the similarities between CR6 and GE15 powders (same chemical composition, same colour, etc.), the direct measure of the composition of a given sample is not possible. Since homogeneity of a mixture is assessed from the variation of the composition between samples, it is important to find an indirect way to measure the composition of an Alumina powders mixture. Therefore, different correlations between measurable parameters (specific surface area, Sauter diameter, cohesion and flow function coefficient) and the composition of the mixtures were found empirically allowing indirect measurement of the composition of a given sample.

The four studied parameters: specific surface area (, measured by nitrogen adsorption using BET method), Sauter diameter (, measured by laser granulometry, dry transportation with air at 0.1 bar), cohesion and flow function coefficient (and , measured by FT4 shear test at 9 kPa on a 10 mL cell), were compared in terms of precision, relevance and feasibility.

It appears that the homogeneity assessment methods involving rheological parameters (cohesion and flow function coefficient) show many advantages compared to the other methods. Not only are they the most accurate correlations found but both parameters are measured simultaneously, meaning that it is possible to apply both methods on the same samples while all the tests (N_2 adsorption, laser granulometry and FT4) are destructive. Those two methods also allow the estimation of the homogeneity at a more interesting scale (around 7 g of powder), indeed, specific surface area and Sauter diameter methods give a homogeneity at a scale of less than one gram which is not relevant for the end-use properties of the final product.

[1] P. Danckwerts, The definition, measurement and some characteristics of mixtures, Applied scientific research. Section A, 3 (1952) 279-296

[2] S. Massol-Chaudeur, H. Berthiaux, S. Muerza, et J. Dodds, A numerical model to identify the structure of a high-dilution powder mixture, *Powder Technol.*, 128 (2002) 131-138

Copyright © American Institute of Chemical Engineers. All rights reserved.

508516 Characterising the Grain Shape: In Search of Size Independent Shape Descriptors

Abstract:

Granular materials contain a large number of grains that differ in size and shape. While the characterisation of the size distribution is typically done using sieve analysis, the characterisation of the shape of the grains remains a difficult task and is one of the major bottlenecks in the characterisation of the granular material. In this study, we attempt to characterise the shape of three different types of grains commonly used in iron and steel making industry, namely pellet, sinter and iron ore lump. While the pellet particles can be considered nearly spherical in shape, imposing the spherical shape on the iron ore lumps and highly irregular sinter particles seems a very poor choice. We consider two different size ranges of pellet particles and four different size ranges for sinter and iron ore lumps. Imaging of the grains is done using an iPhone 7 Plus camera and sufficient number of grains of each type in all the size ranges are considered. We perform the image analysis and obtain various boundary parameters (such as projected area, minor and major axis, perimeter, etc.) of the grains using MATLAB image processing toolbox. In addition, we define two additional boundary parameters to characterise the deviation of the projected area from that of a circle and calculate their values for all the different type of grains considered in this study. Using these length scales, we define a total of nine different shape descriptors and obtain the distribution of these shape descriptors for each type of grain for all the size ranges. Interestingly, for a given type of grains, the frequency distribution curves for these nine shape descriptors turn out to be nearly independent of the size range of the grains. This suggests that the chosen shape descriptors indeed represent properties related to the shape of the grains alone, irrespective of the size of the grains considered.

We investigate the correlation among these shape descriptors by evaluating the values of the correlation coefficients between each pair of the shape descriptors using the Pearson method, Kendall method and Spearman method. Our results indicate that out of the nine shape descriptors considered in this study, five shape descriptors are strongly correlated with each other and only the remaining four shape descriptors need to be considered in characterising the shape of the grains. Out of these four uncorrelated shape descriptors, two shape descriptors, namely the aspect ratio of the grains and the root of form factor, have been considered important by researchers in previous studies. The two other shape descriptors proposed in this study, which are used to characterise the deviation of the projected grain shape from that of a circle, are not measured previously to the best of our knowledge. We compare these four shape descriptors for the pellets, sinter and iron ore lumps having the same sieve size range and show that the shape descriptors for these three different types of grains are different.

We next calculate the values of these four shape descriptors for relatively larger coke particles using 2d image analysis. We also perform 3d image analysis of some of these coke particles and calculate the values of the analogous shape descriptors in 3 dimensions. We investigate if the identified 2d shape descriptors (obtained from 2d imaging) can be correlated to the analogous 3d shape descriptors (obtained from 3d imaging) and explore if it is possible to use the 2d imaging technique to estimate the 3d shape descriptors, thus eliminating the need of 3d imaging.

Copyright © American Institute of Chemical Engineers. All rights reserved.

508543 Investigating Mechanical Properties of Surfactant Films at the Solid-Liquid Interface Using AFM

Abstract:

Corrosion is a well-known problem in the oil and gas sector and also in many other industries. Surfactant based inhibitors have been used for corrosion control in oil pipelines but their inhibition mechanism is still not fully known. Our group is studying surfactant corrosion inhibition mechanism by various characterization techniques (AFM, Electrochemical and Spectroscopic techniques). A better understanding of corrosion inhibition mechanism will lead to the design and development of not only cost-effective but also greener surfactants system for corrosion inhibition purposes.

Atomic force microscopy (AFM) is an effective tool to study the barrier properties of surfactant films at the solid-liquid interface. Determination of mechanical and topographical properties of single and mixed surfactant coatings by atomic force microscopy (AFM) can delineate the role of physical vs. electrochemical barrier by such films in the corrosion inhibition process. . So far, the AFM data, obtained using a closed cell, indicate the effect of surfactant concentration and that of hydrocarbon chains interactions on the mechanical properties of cationic surfactant films. Ultimately, integration of the AFM and electrochemical results will reveal the correlation between the surfactant structures at the solid-liquid interface and their corrosion inhibition efficacy.

Copyright © American Institute of Chemical Engineers. All rights reserved.

508545 Deposition Rate Consequences of the Formation of Multi-Spherule Cluster Aggregates in Gases "Role of Momentum Shielding^{TM}"

Abstract:

For engineering flow systems characterized by particulate suspensions in gases, will the formation of *cluster aggregates* (CAs) in the mainstream accelerate deposition rates on immersed targets or confining walls ---or is aggregate formation more likely to suppress capture rates? Recent research on the drag properties of large cluster aggregates (e.g., stimulated in part by the Sorensen (2011) review) has opened the door to answering these important mechanism- (and pressure-) dependent questions. Sub-100 nm "primary" spherules, which may find themselves in even tenuous 'fractal-like' N-spherule clusters, can take advantage of what we call "momentum shielding"---analogous to the drag advantages experienced by birds who elect to fly in formation. We show here that the extent of this "momentum shielding" can be conveniently quantified (in a dimensionless function we call: $S_{mom}(N;Kn_1;\text{aggregate structure})$), and used to predict the deposition-rate consequences of aggregation in aerosol flow systems when the cluster deposition mechanism is dominated by either: i) isothermal convective diffusion (C-D), ii) thermo-phoresis (TP-) or iii) inertial impaction (II). As expected, we show that for any type of aggregate structure (including 'fractal-like'), momentum shielding increases with cluster size (e.g. N) and gas pressure (via the reciprocal Knudsen number based on spherule radius R_1). We also show how S_{mom} can be calculated for the limiting case of a $N>>1$ random close-packed-$D_f=3$ cluster ---which is nearly gas impermeable. For fractal-like aggregates, including DLCAs ($D_f=1.8$) and RLCAs ($D_f=2.1$) we have recently shown that the abovementioned function: $S_{mom}(N, Kn_1; D_f)$ is actually sufficient to make useful engineering estimates of CA Brownian diffusion coefficients (Rosner and Tandon (2017b), thermophoretic diffusivities (Rosner and Tandon (2017a) and aggregate "stopping times" (Rosner and Tandon (2017c)----each in terms of their previously known values (D_1, $\alpha_T D_1$, and $t_{p,1}$) for isolated individual spherules in the same environment.

This information now enables instructive engineering calculations of the deposition-rate consequences of spherule aggregation to form large cluster aggregates-----i.e. at the same mainstream spherule volume fraction we can now determine how much aggregation will increase or decrease the total spherule acquisition rate on an immersed target. We present here representative results for each of the abovementioned particle transport mechanisms----i.e.: i) isothermal convective-diffusion (C-D), ii) thermo-phoresis (TP-) or iii) inertial impaction (II)----- including a brief account of the consequences of dealing with Kn1-dependent coagulation-'aged' aggregate size distributions (ASDs) in the mainsteam. We summarize our current re-examination of the canonical problem of inertial impaction on a circular cylinder in $Re>>1$ crossflow (Rosner and Tandon (2017c), dealing now with CA-"projectiles" but in the absence of spherule loss from the much larger target. We conclude by briefly outlining our planned work on the efficient "inversion" of sampled cluster-ASD-data to reconstruct, using the abovementioned $S_{mom}(N,Kn_1;D_f$ information, what ASD must have existed in the mainstream---at least in the absence of appreciable supercritical aggregate break-up.

References:

1. Rosner, D. E. and P. Tandon (2017a). Knudsen Transition Effects on the Thermophoretic Properties of Fractal-like Aggregates; Implications for Thermophoretic Sampling of High-Pressure Flames, Aerosol Science and Technology (in press).

2. Rosner, D. E. and P. Tandon (2017b). Aggregation- and Knudsen Transition Rarefaction- Effects on Particle Mass Deposition Rates by Convective-Diffusion, Thermophoresis or Inertial Impaction; Consequences of Multi-spherule 'Momentum Shielding', Aerosol Science and Technology (in press).

3. Rosner, D. E. and P. Tandon (2017b). Mainstream Aggregation and Polydispersity Effects for Inertially-Dominated Deposition Rates on Cylinders (or Leading 'Edges') in Crossflow"---- Consequences of 'Momentum Shielding' for the Impaction of Large, "Compact" ($2<D_f<3$) Multi-spherule Aggregates, Aerosol Science and Technology (prepared for submission, August 2017).

4. Sorensen, C. M. (2011). The Mobility of Fractal Aggregates: A Review, Aerosol Science and Technology. 45: 765-779.

Copyright © American Institute of Chemical Engineers. All rights reserved.

508609 Experimental and Numerical Studies on the Thermal Diffusivity of PackedÂ Powder Beds

Abstract:

Several industrial processes, such as calcination and drying, involve thermal treatment of particles. In order to achieve efficient processes and good quality products, the thermal properties of the particles and that of the bulk powder must be well defined. Previously, some models, such as the Krischer and Maxwell-Eucken models, have been proposed to relate particle thermal properties to that of the bulk powder. Though these models have been shown to work well, it is not very clear how to account for changes in the packing density, moisture content, and arrangement of particles. In addition, the particle thermal properties of most materials are not readily available, which limits the application of these models. To get a better understanding of the relationship between the particle thermal properties and that of the bulk powder and also to quantify the effects of packing density and moisture content on the bulk powder thermal properties, we run several experiments and discrete element method (DEM) based simulations. In both the experiments and simulations, a pipe is filled with particles where the temperature of the pipe is kept constant. We measured the temperature change in the powder bed and compute the actual thermal diffusivity of the powder bed. We tested several types of powders and glass beads in the experiments. The packing fraction, the moisture content, and size of particles were varied in the experiments. In the simulations, in addition to the packing fraction and particle size, the thermal properties of the particles were also varied. In this talk, we will present the details of the experimental and simulation results.

Copyright © American Institute of Chemical Engineers. All rights reserved.

509556 Determination of Size, Size Distribution and Refractive Index of Artificial and Biological Microparticles

Abstract:

For the approval of clinical studies, it is mandatory to validate the production process of artificial oxygen carriers by reliably specifying the properties of the microparticles used like the newly developed blood substitute HbMP-700 (1-3). In particular, besides shape, size and size distribution, the amount of hemoglobin (Hb) and methemoglobin (metHb) content is essential.

We present a method based on the simultaneous measurement of collimated and diffuse transmission spectra in the visible and ultraviolet range and numerical analysis to infer the wavelength-dependent refractive index, size and size distribution of microparticles.

For the validation of the method, we used aqueous suspensions of monodisperse polystyrene microspheres of known concentration, determined by flow cytometric measurements. The related data analysis is carried out for synthetic and experimental data. For homogeneous spheres, Mie theory is applied to compute the solution for the scattering of a plane electromagnetic wave by a particle and to calculate the extinction cross section C_{ext}, i.e. the shadow cast by the particle as a function of the vacuum wavelength λ, the particle size, and the complex refractive index $n(\lambda)$. For dilute suspensions of particles, the measured quantity is an ensemble-averaged extinction cross section C_{ext} . This average is obtained by integrating over the particles' size distribution, which may not be known a priori. From the resulting spectral extinction cross sections, the complex refractive index and the mean and standard deviation of the size distribution can be inferred by means of nonlinear least-squares optimization. In order to reduce the number of free parameters of the problem to a feasible amount, $n(\lambda)$ is represented in a set of harmonic-oscillator-like resonance functions. The refractive index of polystyrene microspheres obtained is in good agreement with literature values for the bulk material, the mean and standard deviation of the distribution of particle *size D* are determined with an uncertainty in the range of a few nanometers. For example, for microspheres of mean diameter mean(D) = 2μm and distribution width std(D) = 9.4nm the uncertainty of the mean is u{mean(D)} = 1.2nm and the uncertainty of the distribution width amounts to u{std(D)} = 0.4nm.

In a second step, we applied the method to characterize HbMP-700 microparticles, approximately featuring a prolate spheroidal shape with 700nm length and 400nm width. Besides shape, size and size distribution, the amount of hemoglobin (Hb) and methemoglobin (metHb) content is essential. The relative concentrations of oxyHb, deoxyHb, metHb are accessible from spectral extinction and diffuse transmission measurements, since the different Hb variants exhibit different wavelength-dependent refractive indices. The method to infer size and refractive index described above can be applied to these non-spherical particles, since the deviation of their light scattering properties from those of spheres turns out to be small. The results based on the optical method will be compared to alternative procedures for HbMP particle characterization with respect to their size and size distribution, i.e. electron microscopy, centrifugation and dynamic light scattering. In addition, x-ray spectrometry will be used to detect the ratio between Fe^{2+} (oxy/dexyHb) and Fe^{3+} (metHb) as well as the iron concentration.

References

1) Novel hemoglobin particles--promising new-generation hemoglobin-based oxygen carriers. Bäumler H, Xiong Y, Liu ZZ, Patzak A, Georgieva R. Artif Organs. 2014 Aug;38(8):708-14. doi: 10.1111/aor.12331.

2) Nonvasoconstrictive hemoglobin particles as oxygen carriers. Xiong Y, Liu ZZ, Georgieva R, Smuda K, Steffen A, Sendeski M, Voigt A, Patzak A, Bäumler H. ACS Nano. 2013 Sep 24;7(9):7454-61. doi: 10.1021/nn402073n.

3) Hemoglobin-based oxygen carrier microparticles: synthesis, properties, and in vitro and in vivo investigations. Xiong Y, Steffen A, Andreas K, Müller S, Sternberg N, Georgieva R, Bäumler H. Biomacromolecules. 2012 Oct 8;13(10):3292-300. doi: 10.1021/bm301085x.

Copyright © American Institute of Chemical Engineers. All rights reserved.

517131 FlowCam NanoÃ,Â® Provides Counts, Sizes and Images of Nano-and Microparticles: Application to a Therapeutic Protein Pumping Study

Abstract:

Sub-visible particle characterization is becoming a more important method in assessing drug substance and drug product stability. However, the current USP protocol does not require any characterization of particles below 10 Î¼m. This greatly hampers our ability to properly assess protein stability as a function of particulate formation and obscures most sub-visible particles and various morphological properties.

In this study, we assess the capabilities of the new FlowCam Nano from Fluid Imaging Technologies, in conjunction with a pumping study that focuses on particulate formation generated during fill operations. Initial experiments show that the new FlowCam Nano can generate highly resolved images of nanoparticulates, allowing the simultaneous assessment of particulate morphology for nano- and microparticulates. This is extremely useful, as currently unpublished data (for the pumping study) show that as a function of post-pumping agitation, a significant increase in microparticle concentration coincides with a significant depletion of the nanoparticle concentration. This suggests that agglomeration (and probably not nucleation) is the more likely mechanism for larger microparticulate formation.

The ability to characterize the nano- and microparticulate morphology simultaneously, will allow us to better assess the cause of particulate formation (such as observing the transition of nanoparticles to microparticles and seeing if specific morphological properties of larger nanoparticulates are conserved in microparticulates). In addition, the FlowCam Nano is also sensitive enough to assess differences in the colloidal stability of a drug product in different formulations. During filling operations, and accelerated degradation studies, FlowCam Nano showed the differences in size distributions of IVIG in PBS and 0.25 M glycine buffer. Since IVIG is less colloidally stable in PBS, the size distribution varies significantly more per size bin and the overall mean effective spherical diameter is also ~ 600 percent greater after 24 hours of agitation in comparison to the size distribution data of IVIG in glycine. Thus, Flow Cam Nano has capabilities that are very useful in assessing drugs in the substance (e.g. optimizing drug formulations) and product (e.g. assessing particulates post fill & finish operations) phases.

Copyright © American Institute of Chemical Engineers. All rights reserved.

Numerical and experimental estimation on the normal and tangential capillary bridge force adhered to two spheres

Kazuo Murase*, Keisuke Arai, Takato Ootsuka, Daiki Sakamoto and Futa Egawa,
Department of Applied Chemistry, Chuo University
1-13-27, Kasuga, Bunkyo-ku, Tokyo,1128551, Japan
**hmurase@kc.chuo-u.ac.jp*

Introduction

Liquid bridges extensively occur as a binder liquid adhered to solid surfaces in a variety of engineering applications such as crystal growth, ink-jet printing, material processing and powder technology. Especially the strength of adhesion and cohesion of liquid bridges adhered to particles contribute to the consolidation of an agglomerate in wet granular materials. Estimation of the attractive or repulsive force between particles is important in controlling kinetics of agglomeration and handling of granular materials. Characteristics of a liquid bridge e.g. the equilibrium bridge geometry [1], stability [2], static and dynamic binding force [3-6], rupture energy [7,8], and wetting hysteresis [9] etc., have been experimentally and numerically investigated.

The liquid distributions within a particles bed are classified into four types of categories, i.e. pendular, funicular, capillary and slurry states, according to the liquid contents [10]. The static normal capillary force at pendular liquid state have elaborately been investigated as Fisher [11], toroidal model [8] and the exact solution of Yung-Laplace equation [12]. The dynamic normal capillary forces [13] are experimentally estimated with lubrication theory. The tangential force of a liquid bridge [14] also contributes to the primary factor for the granular consolidation with the shear stress of wet granular materials but have not yet substantially been investigated.

This work aimed to estimate experimentally and numerically the normal and tangential capillary force of a static and dynamic liquid bridge adhered to two spheres. The asymmetry static liquid bridge with the shear stress demand for the exact solution of Young-Laplace equation for the equilibrium liquid drop including the effect of gravity.

1. Theory

1.1 Formulation of a asymmetry static liquid bridge

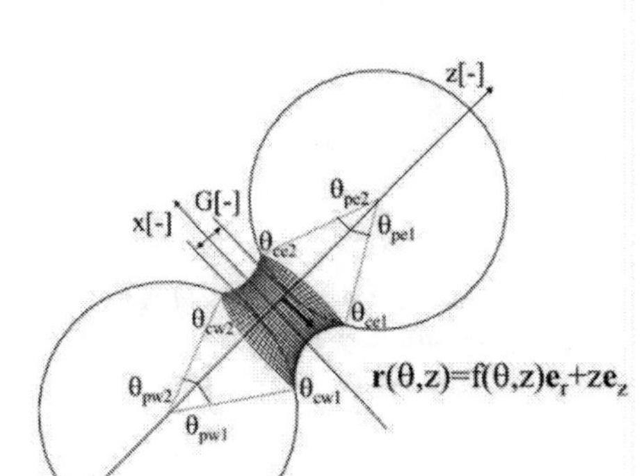

Figure 1 Geometry of a liquid bridge adhered to two spheres with arbitrary inclination in cylindrical coordinate system.

The asymmetry static bridge shape demands for the exact solution of the Young-Laplace equation due to the limit of the toroidal approximation. A prevalent geometry of a liquid bridge adhered to two spheres is drawn in Figure 1. The dimensionless position vector of a liquid surface in cylindrical coordinates is described as $\mathbf{r}(\theta,z) = f(\theta,z)\mathbf{e}_r + z\mathbf{e}_z$ divided by spherical radius, R', which $\mathbf{e}_r$, $\mathbf{e}_z$ are unit basis vectors in r-,z-directions. The symbol (') means the dimensional value. Fig.1 indicates the two dimensional front view with four half-filling (θ_{Pw1}, θ_{Pw2}, θ_{Pe1}, θ_{Pe2}) and contact angles (θ_{Cw1}, θ_{Cw2}, θ_{Ce1}, θ_{Ce2}).

The unit outward normal vector of a bridge surface, $\mathbf{n}_L$, is presented as follows [15],

$$\mathbf{n}_L \equiv \frac{r_\theta \times r_z}{|r_\theta \times r_z|} = \frac{f\mathbf{e_r} - f_\theta \mathbf{e}_\theta - ff_z \mathbf{e}_z}{[f^2(f^2+f_z^{\,2})+f_\theta^{\,2}]^{1/2}} \quad \ldots(1)$$

,where low suffixes (r,θ,z) describe partial derivatives.

The dimensionless effective potential energy, $E=E'/\sigma R'^2$ with Lagrangian multiplier, $K=K'/\sigma R'^2$, on the constrained condition of a constant volume, $V=3V'/4\pi R'^3$, is described as follows [16],

$$E=\int_{-\pi/2}^{\pi/2}\int_{z_{min}}^{z_{max}} 2\left[f^2\left(f^2+f_z^{\,2}\right)+f_\theta^{\,2}\right]^{1/2} d\theta dz+\frac{Bo}{2}\int_{-\pi/2}^{\pi/2}\int_{z_{min}}^{z_{max}}(z\sin\beta+x\cos\beta)f^2 d\theta dz + K\left[V-\frac{1}{2}\int_{-\pi/2}^{\pi/2}\int_{z_{cp1}}^{z_{cp4}} f^2 d\theta dz\right] \quad \ldots(2)$$

Bond number (Bo) is a dimensionless number defined by $(\rho_L-\rho_G)gR'^2/\sigma$, where is the ratio of gravitational force to surface tension force equally as Etovoes number (Eoe).

The integral domain of a bridge surface is determined by the trapezoidal section of the upper and lower surface lines, [$z_{w2}<z<z_{e2}$ at $\theta=\pi/2$, $z_{w1}<z<z_{e1}$ at $-\pi/2$] and contact lines, [$z_{w1}<z<z_{w2}$, $z_{e1}<z<z_{e2}$ at $-\pi/2<\theta<\pi/2$] with four contact points settled by four half-filling angles ($\theta_{Pe1},\theta_{Pe2},\theta_{Pw1},\theta_{Pw2}$) in Fig. 1. Two contact lines on the upper and lower spheres were approximated by ellipse lines with parametric aspect ratios, which pass through respectively pairs of two contact points settled with half filling angles, ($\theta_{Pw1},\theta_{Pw2}$) and ($\theta_{Pe1},\theta_{Pe2}$). Filling angles, ($\theta_{PW}(\theta)$, $\theta_{PE}(\theta)$) on the lower and upper spheres have variable properties along the circumferential direction. The formulation of a static bridge shape at equilibrium state results in the extremum problem, which corresponds to the condition that the first order variation equals to zero, $\delta E=0$.

1.2 Estimation of the capillary force of a liquid bridge

The dimensionless mean curvature (H) and contact angles (θ_c) of a bridge surface are respectively calculated by eqs. (3),(4) once the outline of a liquid bridge is numerically evaluated.

$$-2H = \nabla_s \cdot \mathbf{n}_L = \left[(\alpha^2 + f_\theta^{\,2}) - f(f^2 + f_\theta^{\,2})f_{zz} - f(1+f_z^{\,2})f_{\theta\theta} + 2ff_z f_\theta f_{\theta z}\right]/\alpha^3 \quad \text{...(3)}$$

$$,where\ \ is\ \ \alpha^2 = f^2(1+f_z^2) + f_\theta^2$$

$$\theta_c(\theta) = \cos^{-1}(\mathbf{n}_s \cdot \mathbf{n}_L) \quad \text{...(4)}$$

,where contact angles also depend on the circumferential angle.

According to the Young-Laplace equation, the dimensionless pressure difference, $\Delta P= \Delta P'R'/\sigma$, across a liquid surface is related with the mean curvature of the surface.

$$\Delta P = (-2H) \quad \text{...(5)}$$

The capillary force on a liquid bridge adhered to two spheres must be evaluated only on the contact line as known by the boundary method due to unspecified neck radius of a bridge and asymmetric condition. Dimensionless averaged capillary forces, $F_{Ci} = F'_{Ci}/(2\pi R'\sigma)$ on the upper and lower sphere (i=1,2 respectively correspond to lower, upper sphere) are presented as follows by three terms, which correspond to the surface tension, hydrostatic pressure and buoyancy terms along the longitudinal angle (θ). The sign of the capillary force set to be positive for attractive direction.

$$F_{ci} = \frac{1}{\pi}\int_{-\pi/2}^{\pi/2} \sin(\theta_{Pi})\sin(\theta_{Pi} + \theta_{Ci})d\theta - \frac{1}{2\pi}\int_{-\pi/2}^{\pi/2} \sin^2(\theta_{Pi})\Delta P d\theta \mp \frac{2}{3}\mathrm{Bo}\cdot V_{bi} \quad \text{...(6)}$$

, which V_b is sphere volume submerged in a bridge

Finite Element Method is applied for the solution of the $\delta E=0$ equation. Isoparametric elements of a biquadratic polynomial shape function were selected for the integral domain and unknown radius. The set of non-linear algebraic equations for unknown radius at element nodes and Lagrangian multiplier was resolved by the Newton-Raphson method [9].

2. Experiment

Tensile strength of a static and a dynamic pendular liquid bridge

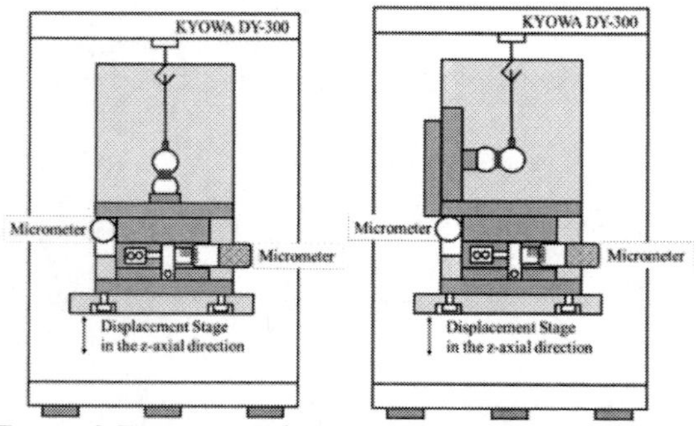

Figure 2 Experimental apparatus for (a) normal and (b) tangential tensile strength of a liquid bridge adhered to two spheres

Figure 2 shows the surface tensiometer (Kyowa interface Science Co., Ltd. DY-300) for measuring (a) the normal and (b) the tangential tensile strength of a water liquid bridge adhered to two stainless steel spheres (radius 3.0mm). The rod attached with one sphere is hanged up to the tensiometer and the other sphere fixed on the displacement stage is moved upward and downward by the tensiometer. The discrete and continuous displacement of the stage are performed for the normal and tangential deformation of a liquid bridge for the static and dynamic states. Two micrometers adjust the appropriate position of the lower spheres for the upper sphere. Water droplets are poured into the spheres aperture and substantially wetted by the displacement of the stage. The tensile strength at static and dynamic state and snapshots of liquid droplets by the digital camera and movie are recorded. The outlines of droplets are evaluated by a digitizer and estimate the half filling angles ($\theta_{Pe1},\theta_{Pe2},\theta_{Pw1},\theta_{Pw2}$) and contact angles ($\theta_{Cw1},\theta_{Cw2},\theta_{Ce1},\theta_{Ce2}$) at four contact points. The sphere configuration (radius, gap), the droplet volume, the half filling angles and the contact lines on spheres initiate parameters of numerical simulation.

3. Results and discussions

3.1 The normal tensile strength of a liquid bridge adhered to two vertical spheres.

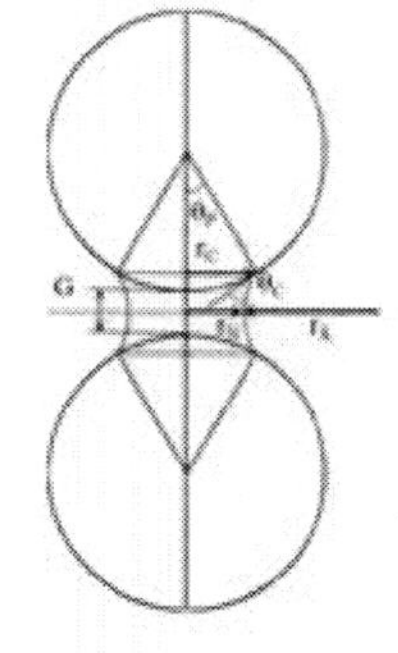

Figure 3 Geometry of toroidal model

The symmetric geometry of a liquid bridge adhered to two vertical spheres without gravity has been extensively investigated by toroidal approximation in Figure 3. The boundary method overestimates the capillary for of a liquid bridge rather than the gorge method but is feasibly and distinctly defined for the asymmetric bridge outline. Symmetric condition simplifies eq.(3) for the mean curvature and eq.(6) for the capillary force into eq.(7),(8) [7,16,17].

$$-2H = \nabla_s \cdot \mathbf{n}_L = \frac{1}{\alpha}\left(\frac{1}{f} - \frac{f_{zz}}{\alpha^2}\right), \text{ where } \alpha^2 = 1 + f_z^2 \qquad \ldots(7)$$

$$F_{Boundary} = r_C \sin(\theta_{Pi} + \theta_G) - \frac{r_C^{\ 2}}{2}(\Delta P) \mp \frac{2Bo}{3} V_{bi}, \quad r_C = \sin\theta_P \qquad \ldots(8a)$$

$$F_{Borge} = r_N - \frac{r_N^{\ 2}}{2} \sin\theta_{cpi}(\Delta P) \mp \frac{2Bo}{3} V_{bi} \qquad \ldots(8b)$$

,where $r_N - r_A = \dfrac{A}{\tan(\pi/2 - \theta_P - \theta_C)} + \sin\theta_P, r_A = \left[A^2 + \{\sin\theta_P - (r_N - r_A)^2\}\right]^{1/2} A = G/2 + 1 - \cos\theta_P$

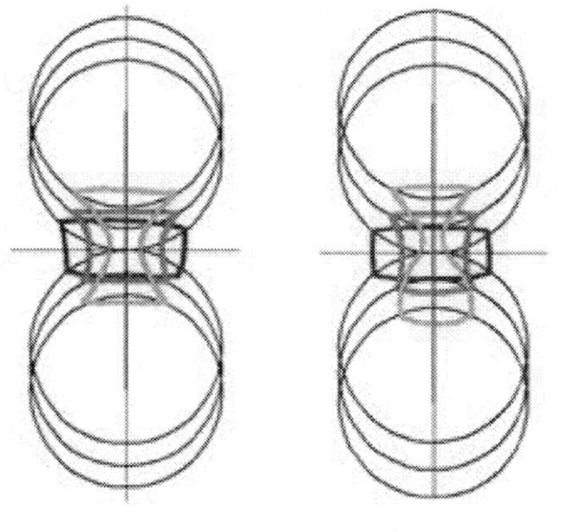

(a) Static state *G=0, 0.47, 0.91* (b) Dynamic state *G=0, 0.59, 1.16*

Figure 4 Experimental droplet outlines at static and dynamic state, Bo=1.21, V=0.088

Figures 4 show symmetric outlines of liquid bridges around the vertical axis respectively at the static and dynamic state. Static state results in the shorter rupture gap and the larger dimensionless normal tensile strength from Figures 5a. Sphere gaps at the peak static capillary force depend on liquid volume. The contact angle on the upper sphere corresponds to the advancing contact angle at the higher Bo number due to gravity effect.

Fig. 4b indicates the downward transition of a liquid immediately before the rupture due to the wettability a liquid. The toroidal approximations over-estimate experimental data at the lower sphere gap but the sphere gaps at the peak dynamic capillary force depend on the velocity of stage displacement. The velocity reduces the peak value of dynamic capillary due to the downward transition of a liquid in Fig. 5b. The stepwise profiles of capillary force are due to the movement sensitivity of tensiometer for the stage displacement.

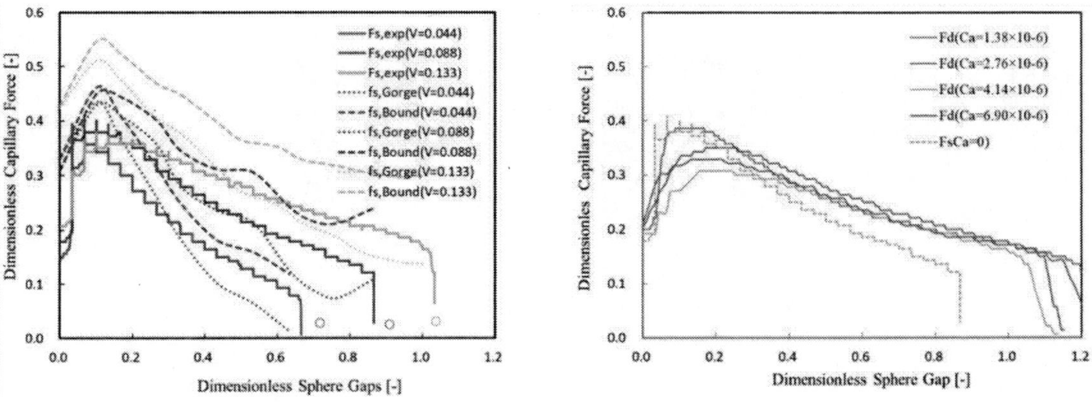

(a) at the static state (b) at the dynamic state

Figure 5 Effect of sphere gap on dimensionless capillary force

3.2 Estimation of boundary condition at contact line by LBM

The Young's equation describes the force balance among three interface tensions (solid-liquid, solid-gas and liquid gas) at three phase contact point and also defines the relationship between interface tension and contact angle. The solution of the Young-Laplace equation at the static state demands for one unknown boundary condition of either a contact angle or a contact line. The simulation for the wettability on three dimensional contact line demands for the elaborate process. Our experiment determines the contact lines on spheres under the constant liquid volume and is utilized for the boundary condition. Besides the perfect wettability the smooth transition of three phase contact lines is unexpected due to the advancing and receding contact angle. Lattice Boltzmann Method (LBM) is expected to estimate the boundary condition at static and dynamic state. The geometry of a liquid bridge is numerically estimated by FEM with half-filling angles predicted by 2 dimensional Shan & Chen model of LBM, which is used by D2Q9 model for transition of distribution function. The estimation for the liquid geometry by FEM has been verified. Figures 6 show the predicted liquid outlines by LBM and FEM. at a given liquid volume.

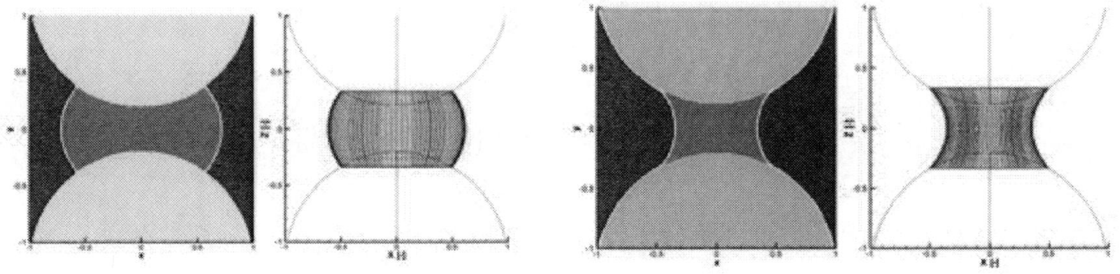

(a) Repulsive interparticle force (b) Attractive interparticle force

Figure 6 Snapshots of liquid geometries at static state by LBM and FEM

Figures 7 show snapshots of dynamic liquid outlines at a given liquid volume up to the rapture state by LBM. The transition of boundary conditions, i.e. filling and contact angles, is smoothly described. Gravity and imperfect wettability, i.e. advancing and receding contact angles, are excluded in Figs. 7 but the effect of gravity on the liquid outline is avairable.

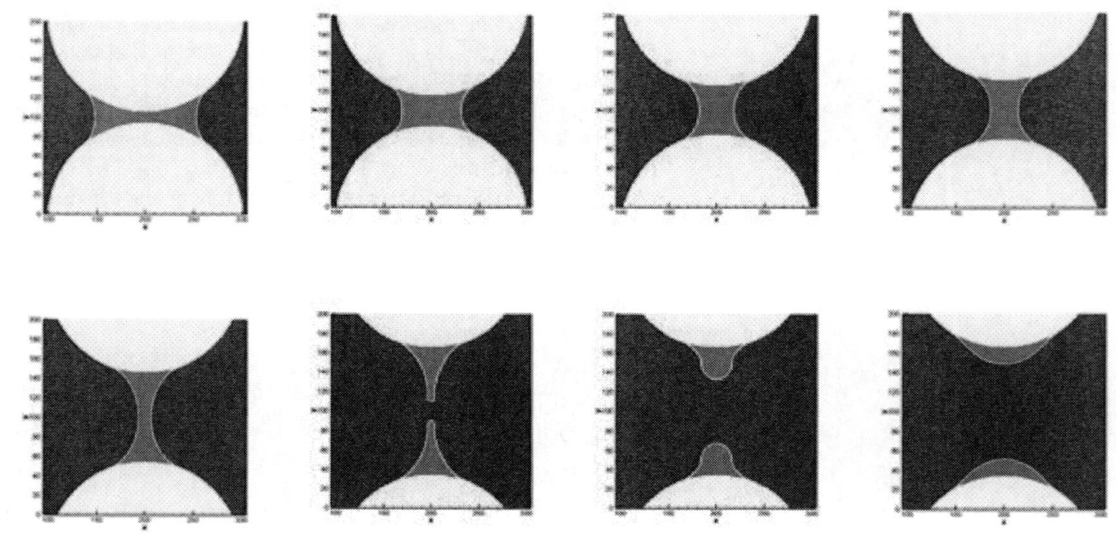

Figure 7 Snapshots of snapshots of dynamic liquid outlines at a given liquid volume

3.3 The tensile strength of a liquid bridge with shear deformation.

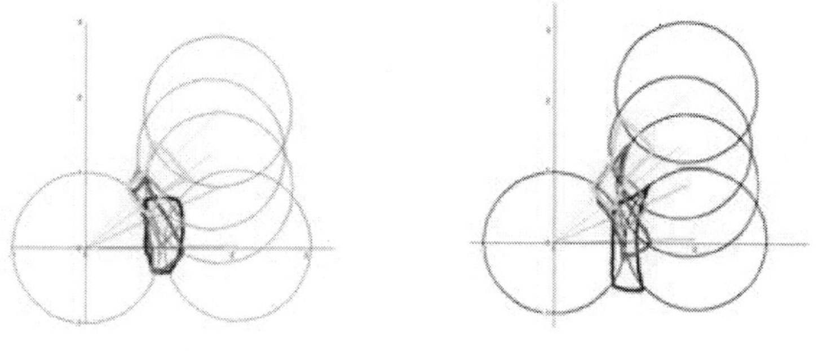

(a) Static state (G=0, 0.59, 0.98) (b) Dynamic state (G=0, 0.59, 0.98)

Figure 8 Liquid outlines at static and dynamic state (Bo=1.21, V=0.088)

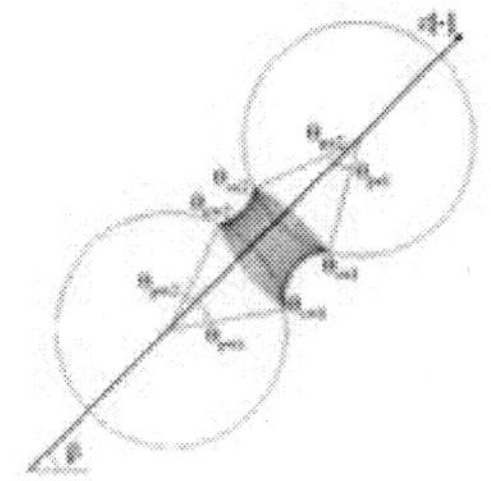

Definitions of half-filling and contact angles.

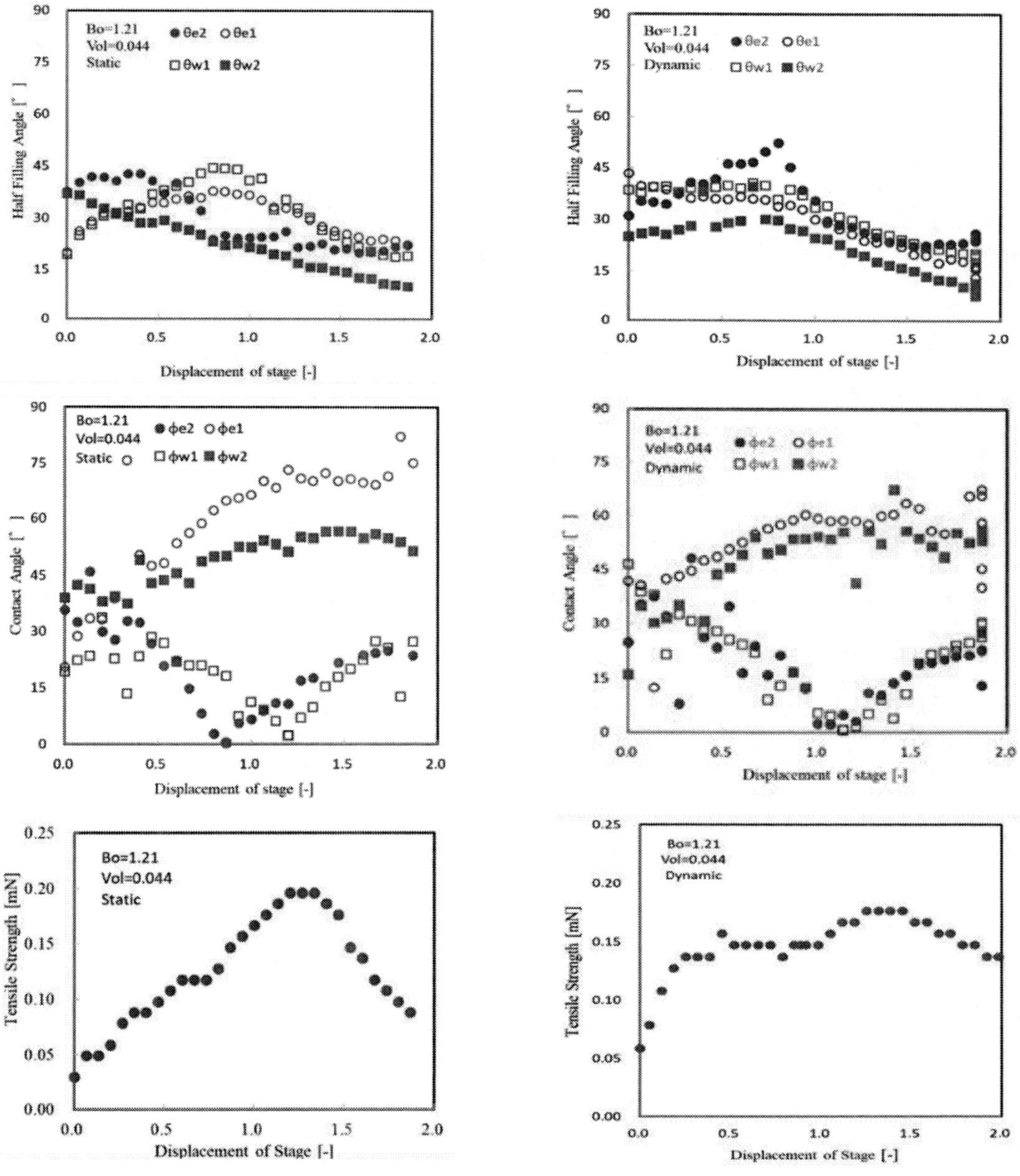

(a) Static state (b) Dynamic state

Figure 9 Profiles of half-filling angles, contact angles and tensile strength , Bo=1.21, V=0.088

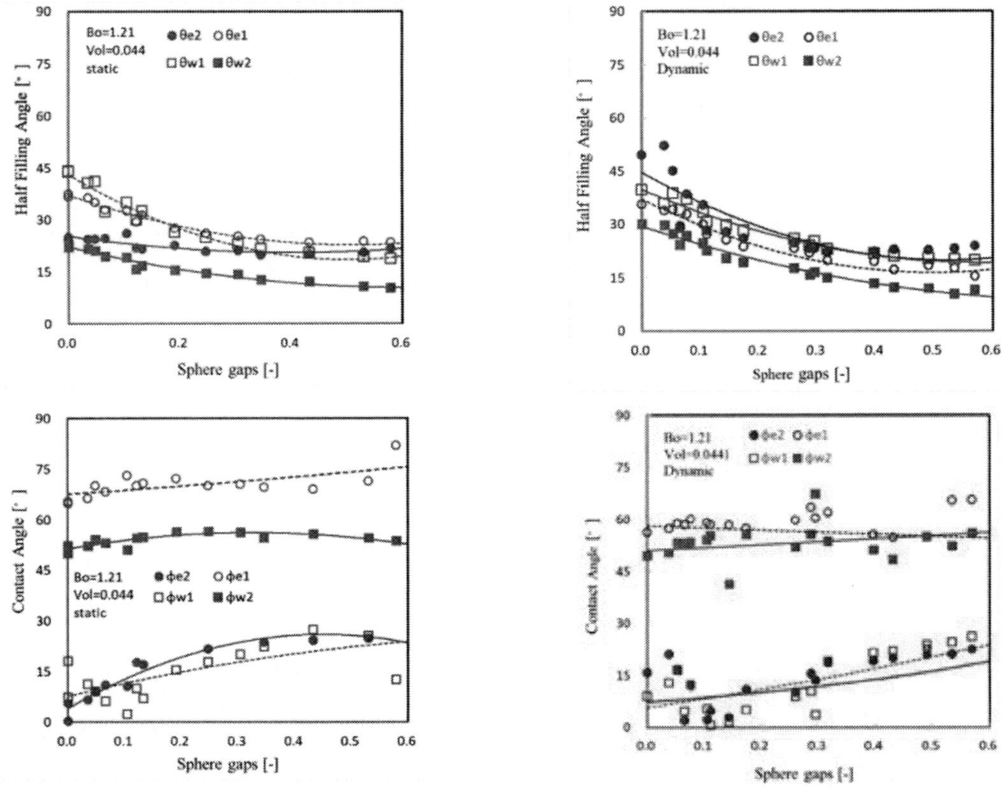

(a) Static state (b) Dynamic state

Figure 10 Profiles of half-filling angles, contact angles with sphere gaps , Bo=1.21,V=0.088

Figures 8 show four types of liquid outlines at static and dynamic state. Magenta and green plots are drawn respectively at the extreme values of half filling angles, contact angles and tensile strengths in Figures 9. During the lower stage displacement two spheres keep at the contact state and the half-filling angles reached to the peak values. Half-filling angles sifted with displacements of a stage have the peak values except the static half-filling angle at the top point of the lower sphere (θ_{Pw2}), which monotonously decrease with stage displacement. The increase and decrease of contact angle respectively corresponds to the advancing and receding contact angle. The contact angles at the bottom point of the lower sphere (θ_{Cw1}) and at the top point of the upper sphere (θ_{Ce2}) transit from the receding to the advancing contact angle due to the infinitesimal values. Figures 10 redraw the half-filling and contact angles with respect to sphere gap. Contact angles contribute to the peak tensile strength, which is measured as resultant force (F_T) of normal (F_n) and tangential (F_t) tensile strength in Figure 11. The normal tensile strength conventionally corresponds to the capillary force. The stage displacement at the peak tensile strength minimize the sum of half-filling and contact angle due to the positive hydrostatic pressure across liquid surface. The larger dynamic tangential strength occurs during the sphere contacting due to the asymmetric deformation. The adhesion force of liquid to sphere at the static state contribute to the larger peak tensile strength.

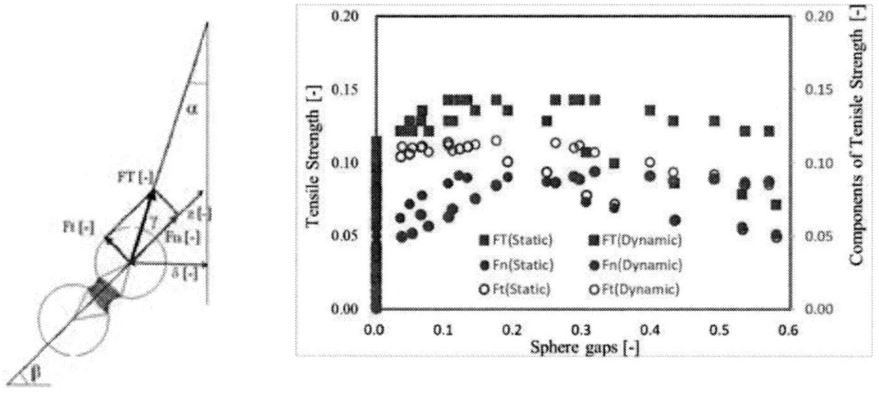

Fig.11 Normal and tangential components of tensile strength at static state

3.4 Numerical simulation at the static state

The solution of Young-Laplace equation demands for one unknown necessary condition on the contact point, i.e. half-filling or contact angle. Liquid surfaces simulated by FEM are drawn with solid lines in Figures 12. The simulation reveals the asymmetric liquid geometry with half-filling and contact angles along longitudinal angle. Disagreements of liquid interfaces results from the approximation of contact lines on spheres. The tangential strength influences depend on the mean curvature. Especially the second figure at the peak half-filling angle has the negative hydrostaic pressure in the direction of relative sphere displacement and positive pressure in the oppsite direction.

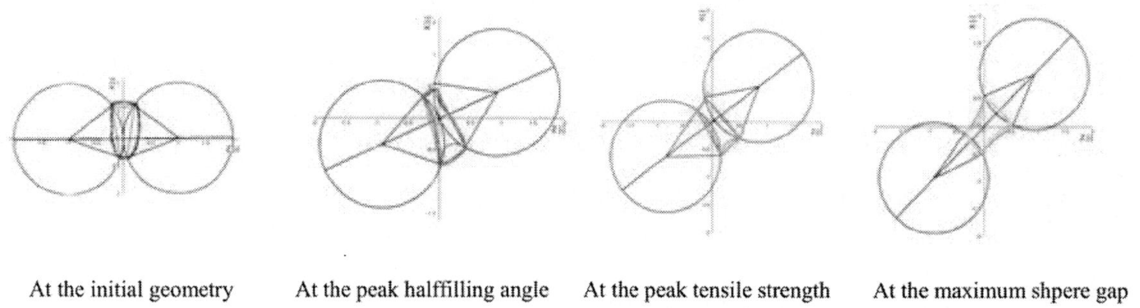

Figure 12 Numerical estimation at the static state

Figure 13 displays transitions of measured normal tensile strengths ($F_{exp,n}$) and simulated capillary force (F_{cal}) to sphere gap. The initial droplet is attached to the above aperture in the direction of sphere movement due to gravity. The stage displacement drifts parts of liquid volume into the lower aperture of spheres. The sphere on displacement stage keep contacting with the other sphere hang up to the tensiometer up to the peak tensile strength. Fcal and FcalB are numerically estimated by outlines of liquid surfaces respectively with Young-Laplace equation and boundary toroidal model. The sign of hydrostatic pressure with deformation of liquid droplets results in the repulsive capillary force. The normal tensile strength corresponds to the capillary force during the range of symmetric liquid bridge. The angle of beta is defined by the inclination of center distance of a sphere pair. The lower normal tensile strength than the capillary force results from the orthogonal component of capillary force at the low range of beta.

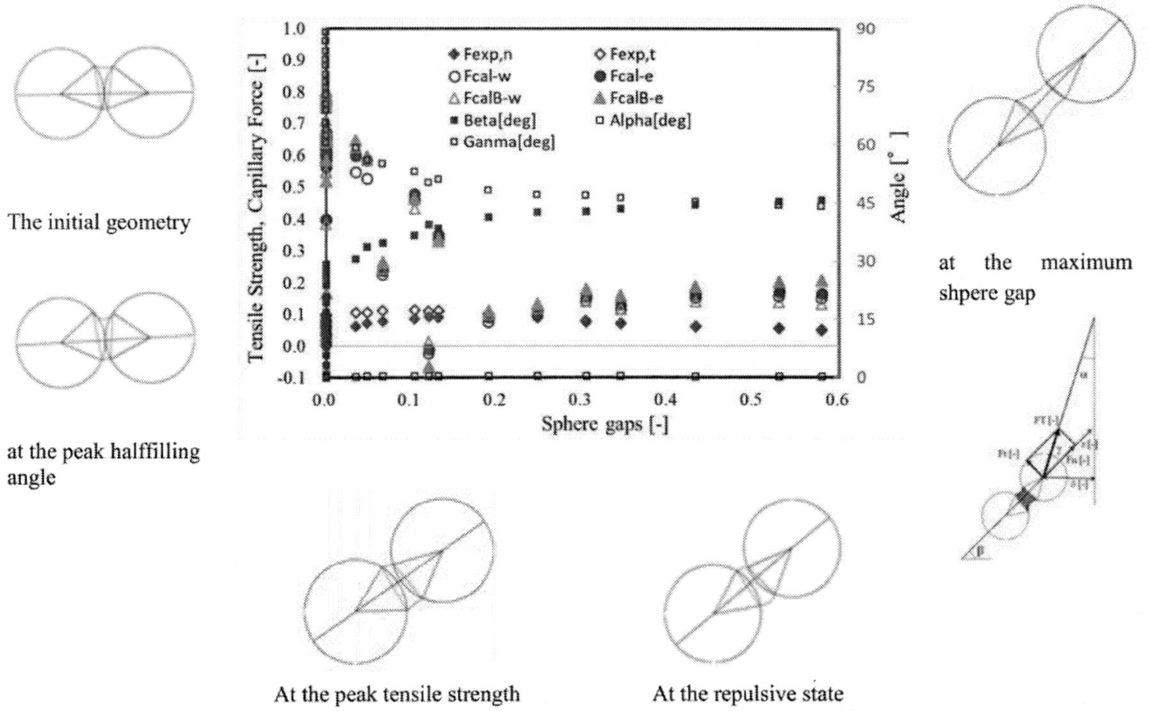

Figure 13 Normal tensile strength and numerical capillary force

Conclusions

Liquid bridges adhered to particles are deformed with the normal and tangential stress in wet granulator. The interparticle force with asymmetric droplet by shear stress must be estimated by the three dimensional solution of Young-Laplace equation. A surface tensiometer and numerical simulation contribute to the normal and tangential tensile strength of liquid bridge adhered to two spheres.

1. The normal strength of symmetric liquid bridge with a surface tensiometer is verified by the conventional toroidal model and the solution of Young-Laplace equation.
2. Sphere gaps at the peak capillary force in the vertical sphere configuration depend on liquid volume and displacement velocity of sphere.
3. The lattice Boltzmann Method is expected to numerically predict the boundary conditions on the contact lines, which are half-filling angle and contact angle at the static and dynamic state.
4. The tensile strength depends on transition of contact angles, which correspond to advancing or receding angles. The larger sum of contact and half-filling angles than the orthogonal angle corresponds to the positive hydrostatic pressure across liquid surface.
5. The normal tensile strength is related with the solution of Young-Laplace equation by FEM. The tangential tensile strength depends on the liquid transition within droplets and the variable mean curvature.

References

[1] K.Murase, H.Shibata, Y.Yasuda and S.Toyama, *Jpn J. Multiphase Flow,* **5**, 322(1991)

[2] A.E.Saez and R.G.Carbonell, *J. Colloid Interface Sci.*, **140**, 408(1990)

[3] D.N.Mazzone, G.I.Tardos and R.Peffer, *Powder Tech.*, **51**, 71(1987)

[4] C.D.Willet, M.J.Adama, S.A.Johnson and J.P.K.Seville, *Langmuir*, **16**, 9405(2000)

[5] M.J.Adama, S.A.Johnson, J.P.K.Seville and C.D.Willet, *Langmuir*, **18**, 6180(2002)

[6] R.K.Niven, *J Petroleum Sci. Eng.*, **52**, 1(2006)

[7] O.Pitois, P.Moucheront and X.Chateau, *Eur. Phys. J.B*, **23**, 79(2001)

[8] S.J.R.Simons, J.P.K. Seville and M.J. Adams, *Chem. Eng. Sci.*, **49**, 2331(1994)

[9] C.D.Willet, M.J.Adama,S.A.Johnson and J.P.K.Seville, , *Powder Tech.*, **130**, 63(2003)

[10] S.M.Iveson, J.A.Beathe and N.W.Page, *Powder Tech.*, **127**, 149(2002)

[11] R.A.Fisher, *J. Agri. Sci.*, 16,492(1926)

[12] F.M.Orr, L.E.Scriven and A.P.Rivas, *J. Fluid Mech.*, **67**, 723(1975)

[13] O.Pitois, P.Moucheront and X.Chateau, *J. Colloid Interface Sci.*, **231**, 26(2000)

[14] U.El Shamy and T.Groeger, *Int. J. Numer. Anal. Meth. Geomeh.*, **32**, 1763(2008)

[15] F.M.Orr, R.A.Brown and L.E.Scriven, *J. Colloid Interface Sci.*, **60**, 137(1977)

[16] K.Murase, T.Mochida, Y.Sagawa and H.Sugama, *Adv. PowderTech.*, **19**, 349(2008)

[17] C.D.Willet, M.J.Adama,S.A.Johnson and J.P.K.Seville, *Handbook of powder technology*, **11**, chap.28, ed. A.D.Salman, M.J.Hounslow and J.P.K.Seville,Elsevier (2007)

NUMERICAL SIMULATION OF WIRE-PLATE ELECTROSTATIC PRECIPITATOR - EFFECT OF PARTICLE CONCENTRATION

Ding Yang[1], *Baoyu Guo*[2*], *Jun Guo*[1], *Aibing Yu*[2]

[1] *Fujian Longking Co. Ltd, Longyan, Fujian, 364000, China*

[2] *Department of Chemical Engineering, Monash University, Clayton VIC 3800, Australia*

Introduction

Electrostatic Precipitators (ESP) have been widely used to separate fine particles from large scale industrial exhaust gases, such as coal fired power plants. In-depth knowledge and careful design optimization are critical to meeting the increasingly strict emission standard in terms of $PM_{2.5}$ emission limit, e.g., 5-10 mg/Nm^3. Numerical modelling has been employed as a powerful tool in this area. Complicated interactions among electric field, gas flow and particle transport are summarized by Adamiak [1]. To model such a system, various levels of simplifications have been made in published papers.

Among them, Lagrangian particle tracking technique is a widely accepted approach. The mass concentration of ash particles in coal-fire flue gas is typically 10-50 g/Nm^3, corresponding to volume fraction in the order of 10^{-5}. Such a small value is often regarded as sufficiently dilute particle concentration in numerical modeling of multiphase flows. For the case of ESP, complicated interactions among the e-field, gas flow and charged particles are involved, even if the deposited dust layer is ignored. Full coupling between the continuous gas phase and discrete particle phase is computationally expensive to achieve a converged steady solution. To avoid such a high cost, the model for ESP was simplified as so called one-way coupling in many previous works, so that the electric field is solved without the particle phase (Guo et al. [2,3]; Farnoosh et al. [4]). This simplification may be acceptable in the end stage of ESP, but will introduce significant error in the early collection stage. Limited studies reported in the literature have considered the particle charges as a source term in the Poisson's equation of electrostatics. For example, Choi and Fletcher [5] predicted a reduced current for the particle-loaded case (d_p = 1-5μm) compared with the particle-free case, but the corona ion charge density somehow shows an increasing trend along the stream. When fine dust particles are treated, the large specific surface area of the particle phase, combined with reduced collection efficiency, implies high space charge density in the collection region. Its effect was understood as a partial suppression of the corona current [6]. As was experimentally proved by Podlinski et al. [7], space charge of charged particles can contribute to the EHD flow pattern and particle transport. In present work, particle space charge is considered through two-way coupling and interesting results are generated in terms of ESP performance as particle loading changes.

Integrated ESP Modelling

Mathematical Equation

Coupled electric field, gas-particle flow and particle charging processes are simulated using

Eulerian-Lagrangian method in Computational Fluid Dynamics (CFD). Major governing equations of vector format are listed below for completeness. For electric field, Poisson equation of potential

$$\nabla^2 V = -(\rho_{ion} + \rho_c)/\varepsilon_0 \quad \mathbf{E} = -\nabla V \tag{1}$$

Current continuity equation

$$\nabla \bullet ((b\mathbf{E} + \mathbf{U}_f)\,\rho_{ion} - D\nabla\rho_{ion}) = S_c \tag{2}$$

where ρ_{ion} is corona charge density, b the ion mobility, D is ion diffusivity, ρ_c is particle space charge density, evaluated by averaging the charges of particles that pass through a control volume (cell),

$$\rho_c = \frac{1}{V_{cell}} \sum_{n=1}^{N} \dot{n}_p q_p \Delta t \tag{3}$$

Where $\dot{n}_p$ is the number flow rate of a parcel of particles tracked, Δt is the residence time in the cell. The source term, S_c, in Equation (2) is transfer rate of ion charges to particles, which is not considered at the moment, i.e., $S_c = 0$. For the gas flow, the momentum equation is,

$$\begin{aligned} \nabla \bullet (\rho_g \mathbf{U}_g \otimes \mathbf{U}_g) &= \nabla(p + \tfrac{2}{3}\rho_g k) + \nabla \bullet [(\mu + \mu_t)(\nabla \mathbf{U}_g + (\nabla \mathbf{U}_g)^{\mathrm{T}})] + \rho_{ion}\mathbf{E} \\ &- \frac{3\pi n_p d_p \mu}{\mathrm{C_C}}(1 + \mathrm{Re}_p^{0.687})(\mathbf{U}_g - \mathbf{U}_p) \end{aligned} \tag{4}$$

For the particle flow, the force balance of a single particle is

$$m_p \frac{d\mathbf{U}_\mathrm{p}}{dt} = \frac{3\pi n_p d_p \mu}{\mathrm{C_C}}(1 + \mathrm{Re}_p^{0.687})(\mathbf{U}_g - \mathbf{U}_p) + \tfrac{1}{6}\pi d_p^3 \mathbf{g}(\rho_p - \rho_g) + n_e q_e \mathbf{E} + \mathbf{F}_{dis} \tag{5}$$

k-ω SST turbulence model is chosen for gas flow, and the dimensionless Lawless charging rate equation is used for particle charging [8].

Model Conditions

An integrated ESP model developed comprises two parallel modules or submodels, namely, electric field model and coupled gas-particle flow model. One-way coupling between the two modules is used earlier is upgraded to two-way in the present work. The electrostatic field caused by charged particles in suspension is considered in the solution of electric field. The two submodels are coupled through a small under-relaxation factor, in views of strong interaction between the electric field and particles flow, and the stochastic nature of the discrete particle phase. In this way all the volume averaged quantities related to the particle phase (such as the particle space charge density) are smoothed out over time in the simulation.

A single channel, 4-wire and flat plate configuration is considered. The dimensions are 3.5mm in wire diameter, 0.24m for wire-wire spacing, 0.2m for wire-plate spacing. The plate length is 1.2m. The simulation domain is defined as 1.74m long by 0.4m wide box with the 4 holes inside. The grid is smoothly refined at places where higher gradient in electric quantities is expected, such as the proximity of the wires and middle of the adjacent wires. Other base case conditions are as follows.

- Applied voltage on the wire electrode: Va = -60kV. The charge density on the wire surface is determined using an injection law combined with Peek equation (Guo et al.[9]).
- Gas temperature: 120°C. As the inlet boundary condition for the gas phase, the velocity is set as 1 Nm/s; relative turbulence intensity 3.7%, turbulent length scale $0.3D_h = 0.24$m.
- Dust particle size is mono-dispersed at d_p=2.5 μm. Particle density 2000 kg/m^3; ash dielectric constant of κ_p=6.6, Particle loading at the inlet, $(\rho_p)_{in}$, is varied with default value of 10 g/Nm^3.

Typical Results

Electric Field without Particles

When the inlet/outlet boundaries are moved sufficient long distance away from the electrodes in the simulation domain (referred to as "natural boundary condition"), the electric field around the boundary wires show an asymmetric distribution pattern that differ from that around the inner wires. Electric field-lines/current can go out of the limit of the collection region (Figure 1). Electric field intensity on the collection plate is of interest as it is considered a factor to determine the upper limit of operation before spark-over [10]. Normally the field intensity on the plate is highest at the point that faces exactly the wire center with the least distance. This maximum field intensity on the plate is 3.80×10^5 V/m. However, when the natural boundary condition is used, the highest value (1.06×10^6 V/m) predicted in field intensity appears at the endpoint of the plate due to its sharpness. This is a likely point that would trigger electric spark-over prematurely. Practically such a sharp edge should be avoided. Generally, the field strength at the middle wires is smaller than that at a boundary wire, due to the counteraction of adjacent wires. Then the corona charge density at the boundary wires should be greater than that at the inner wires (Figure 2), which is similar to the simulation result of Kasdi [11].

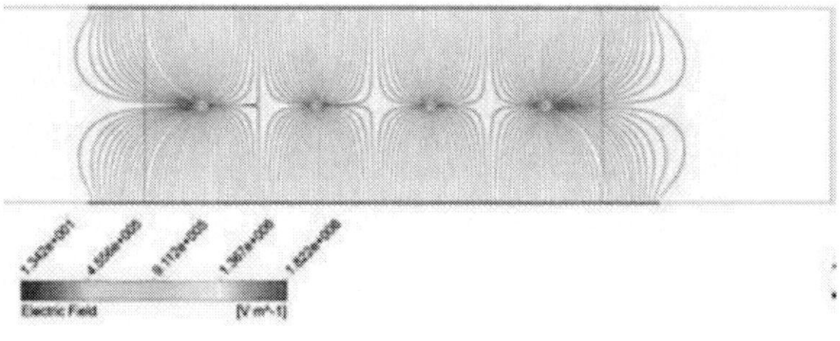

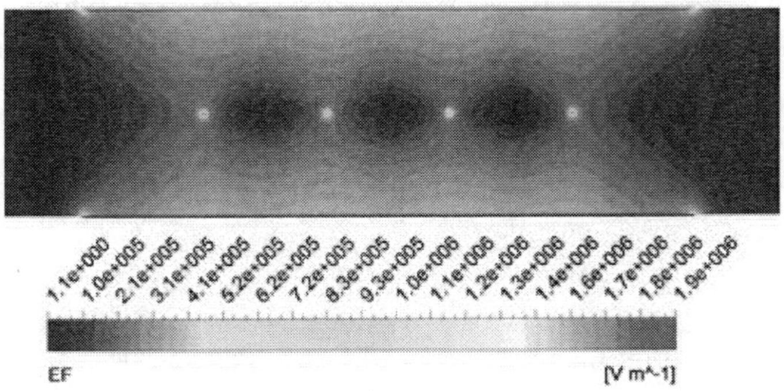

Figure 1. Field streamlines and intensity distribution (no particles).

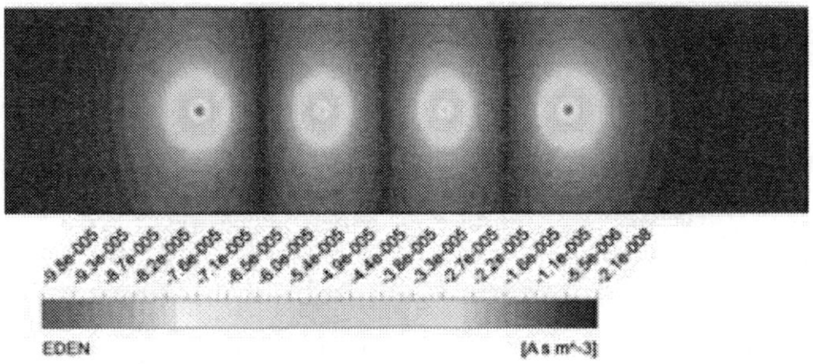

Figure 2. Corona charge density (no particles).

Effect of Particle Space Charge

Particles are randomly injected from the inlet plane with uniform probability in locations and no-slip for velocity (same as the gas velocity). In using random number generator, it is ensured that the injection points are not repeated for every coupling cycle. Figure 3 shows sampled particle trajectories with color scaled to particle charge. Figure 4 shows a pattern of particle deposition rate on the plate. It is seen that the end points of the plate, particularly the downstream one, are special spots to accumulate more dust.

Figure 3. Particle trajectories with colour scaled to particle charge.

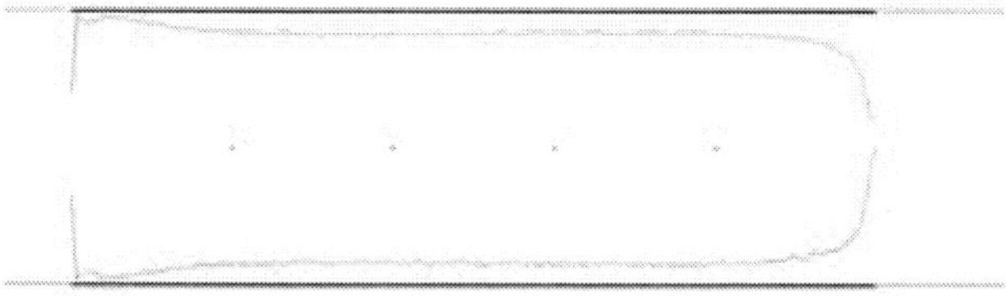

Figure 4. Indicative pattern of particle deposition rate on the plate.

Figure 5 shows the averaged space density of charges carried by particles. Highest space charge density appears near the first wire electrode, where electric current density is highest. It diffuses gradually from the centre line with particles drifting toward the plate while moving forward. Due to the radiating nature of the field-lines around each wire, the lateral component of Coulomb's force in the region close to the centre line may not be as large as that off the centre line. Not collected charged particles of negative polarity are conveyed towards the outlet, lowering the potential downstream outside the collection region (Figure 6), causing distortion of the overall electric field distribution.

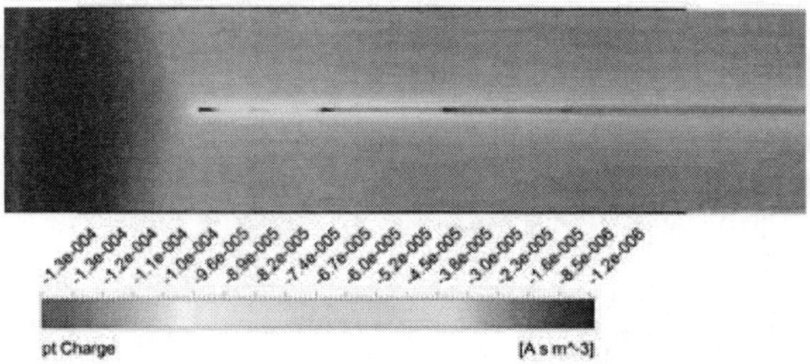

Figure 5. Distribution of space charge density of particles.

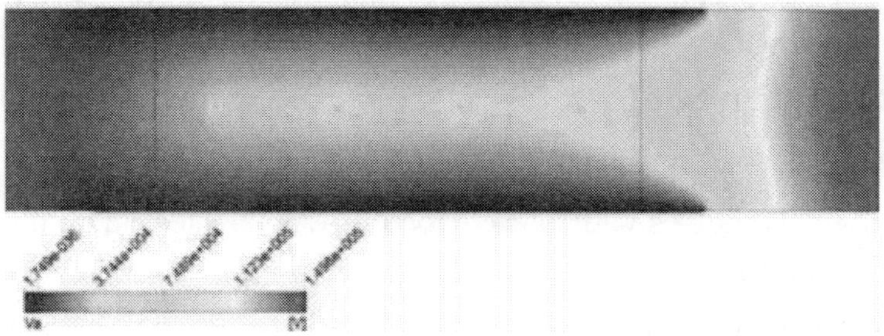
Figure 6. Distribution of electric potential (in absolute value).

Effect of Particle Loading

On Electric Current

Figure 7(a) shows the mean corona current, based on unit length of wire. As can be seen, the corona current drops quickly as particle loading increases, although a fixed voltage is applied. This is caused by the secondary electrostatic field created by the space charges of particles that cancels out the primary electric field at the wire surface and subsequently suppressed the corona generation.

On Particle Collection Efficiency

Predicted particle collection efficiency, as particle loading changes, is given in Figure 7(b). The collection efficiency, increases initially with particle loading (<3 g/m^3), peaks at particle loading of 4-5 g/m^3 and then decreases as particle loading increases further. When ignoring particle space charge, the electric field would be independent of particle loading. This is the case for very low particle loading, where the fractional collection efficiency is constant, or in other words, the quantity of dust collected as well as that penetrated increases proportionally with particle loading.

The observed enhancement of collection by a slight amount of particles is due to the redistribution of electric field so that the near-plate and volume-averaged field intensity happens to increase, which favor the charging of most particles. The asymptotic lower limit of the curve should correspond to the situation when the corona current is so small that turbulent dispersion dominates over the drift force.

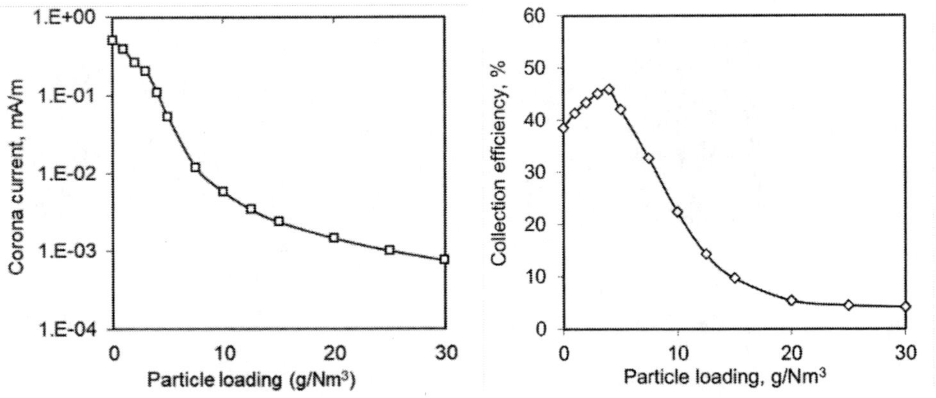

(a) (b)

Figure 7. Mean corona current (a) and collection efficiency (b) as function of particle loading.

On Electric Field Distribution

Figure 8 visualizes the predicted corona charge density and field-lines, which shows how the electric field distribution evolves with particle loading. As the electric field is distorted starting first from the end side, corona density at 3 of the 4 downstream wires is suppressed faster than the 1st wire. When the particle loading is above 7.5 g/m^3, all wires except the first one stop emitting. Meanwhile, reversed field-lines appear around the last wire, and the affected region expands as the loading increases further. Otherwise the field-lines become smoother with increasing loading. Since the field-lines are extended to the outlet, some escaped particles can be pulled back into the collection region, a factor that contributes to significantly greater deposition rate around the end-point of the plate.

Overall, the particle charges affect the particle collection adversely, as they tend to suppress the corona emission or reduce the corona current. Particularly the penetrated particles, which carried significant charges, have prolonged effect as they float along. The corona at the downstream electrodes even extinguish unless the applied voltage is raised individually. When highly charged, more particles precipitate to the plate, reducing the number of particles floating ahead (or less space charge density). The charges carried by these remaining particles exert a negative feedback to the electric field and particle charging, which compromises the collection efficiency.

Electric Quenching

Corona quenching is a common phenomenon in ESP when fine particles are treated. Riebel et al. [12] experimentally studied the corona discharge characteristics of ESP with glycerol aerosol, which found corona onset voltage increases and the corona current significantly decreases after the glycerol steam was injected into the system. Under fixed voltage in the present simulation, corona currents from all the wire electrodes drop with increasing solid loading. The downstream coronas (from the second one) will eventually extinguish at some point. An interesting phenomenon arises when the particle loading increases above 10 g/Nm^3, i.e., the electric streamlines in a small zone around the last wire reverse direction locally. This implies that a charged particle, once falls accidently within the zone, could be trapped and attracted to the same wire, as seen from the particle trajectories in Figure 9. Thus this wire is subject to heavier dust deposition.

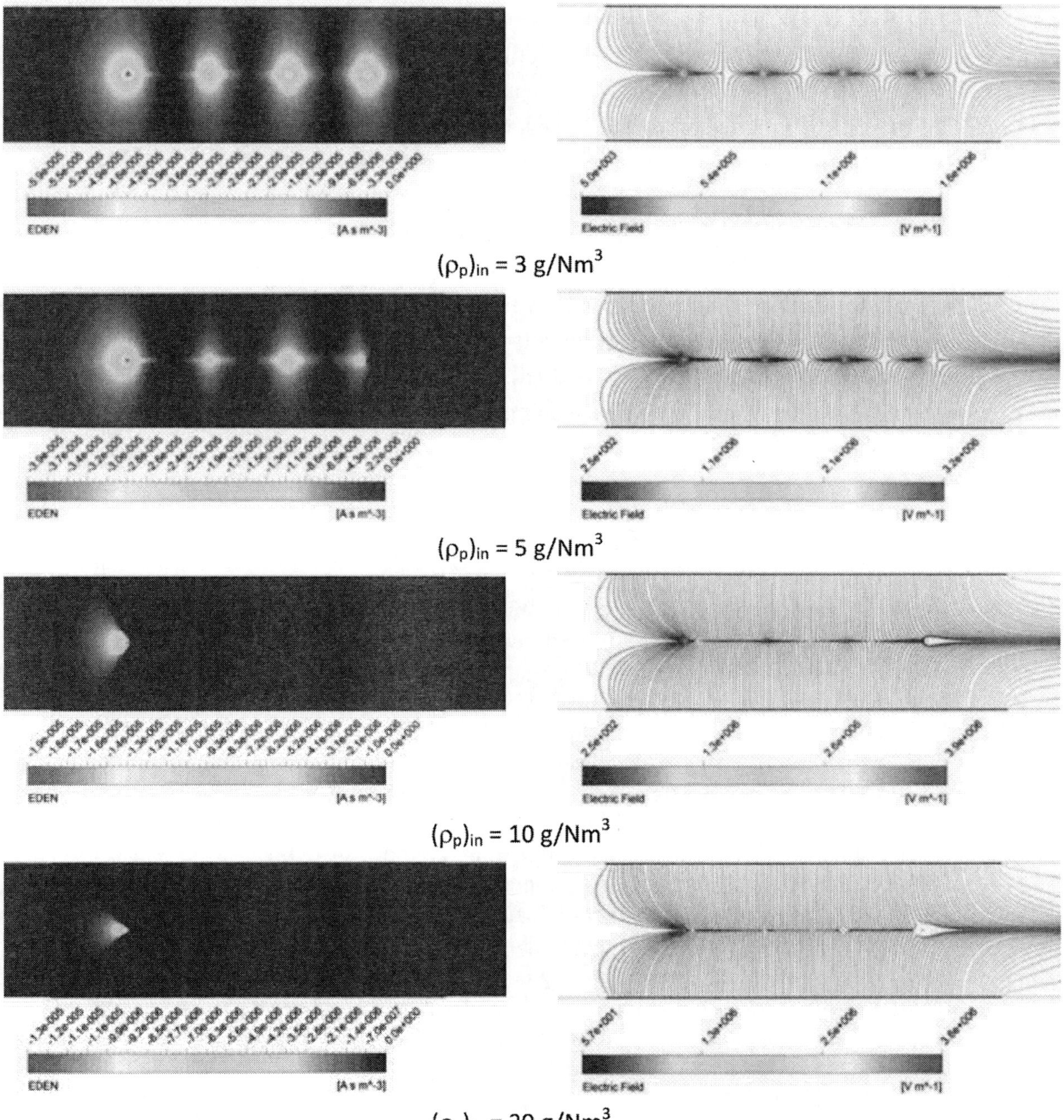

Figure 8. Visualizations of corona charge density and field streamlines for different particle loadings.

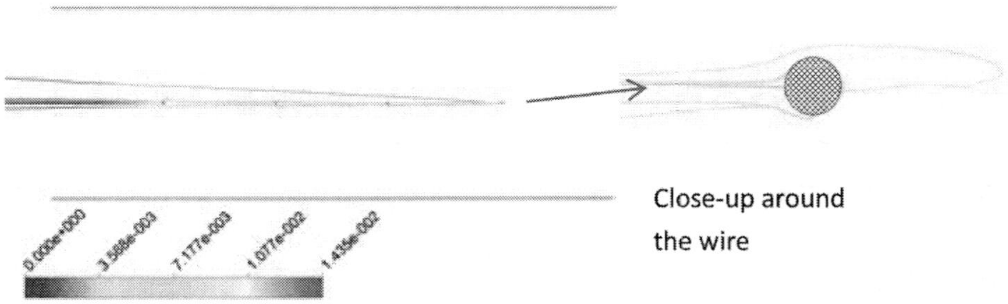

Figure 9. Trajectories of particles falling on the corona wire ($(\rho_p)_{in}$ = 20 g/Nm^3).

Concluding Remarks

CFD based ESP model has been advanced, with an enhanced coupling between electric field and particle flow. So the model can be readily applied to the early collection stage of the ESP, where dust particle loading is too high to neglect. Space charge of particles is considered, and its significance is clarified in a broad range of particle mass concentration in the gas. Dust loading significantly affects the dust collection efficiency for fine particles considered. Maximum collection efficiency exists at a critical particle loading that depends on particle size. However, as particle loading increases, collection efficiency decreases and eventually approaches asymptotically to a lower limit. The phenomenon of electrical quenching is observed, where downstream corona is extinguished or one of the emitters is polluted when the particle loading is high.

References

1. Adamiak K. (2013), "Numerical models in simulating wire-plate electrostatic precipitators: A review", *J. Electrostatics*, 71, pp.673-680.
2. Guo B.Y., Yu A.B. and Guo J. (2014), "Numerical modelling of electrostatic precipitation: effect of gas temperature", *J. Aerosol Science*, 77, pp.102-115.
3. Guo B.Y., Yu A.B. and Guo J. (2015), "Numerical modelling of ESP for design optimization", *Procedia Engineering*, 102, pp.1366-72.
4. N. Farnoosh, K. Adamiak, G.S.P. Castle (2010), "3-D numerical analysis of EHD turbulent flow and mono-disperse charged particle transport and collection in a wire-plate ESP", *J. Electrostatics*, 68, pp.513-522.
5. Choi B.S. and Fletcher C.A.J. (1998), "Turbulent particle dispersion in an electrostatic precipitator," *Applied Mathematical Modelling*, 22, pp.1009-1021.
6. Parker K.R., Applied Electrostatic Precipitation, Blackie Academic & Professional, 1997.
7. Podlinski J., Niewulis A., Mizeraczyk J. and Atten P. (2008), "ESP performance for various dust densities", *J. Electrostatics*, 66, pp.246-253.
8. Lawless P.A. (1996), "Particle charging bounds, symmetry relations, and an analytic charging rate model for the continuum regime", *J. Aerosol Sci.*, 27, pp.191-215.
9. Guo B.Y., Guo J. and Yu A.B. (2014), "Simulation of the electric field in wire-plate type electrostatic precipitators", J. Electrostatics, 72, pp.301-310
10. White H.J. (1974), "Resistivity problems in electrostatic precipitation", *J. Air Pollut. Control Assoc.* 24, pp.313–338.
11. Kasdi A. (2016), "Computation and measurement of corona current density and V–I characteristics in wires-to-plates electrostatic precipitator", *J. Electrostatics*, 81, 1-8.
12. Riebel U., Radtke R. and Loos R. (2002), "An experimental investigation on corona quenching", *J. Electrostatics*, 54, pp.159-165.

Triboelectric Charge of Spherical Glass Particles against Metal Pipeline

HoSu CHOI [1], Teruo SUZUKI [2] and Kwangseok CHOI [3,]*
[1] Christian Academy in Japan, [2] Kasuga Denki Inc.,
[3] National Institute of Occupational Safety and Health, Japan
**choiks@jniosh.go.jp*

Introduction

Insulating powders are used in many ways such as in the food, pharmacy, and chemical industries. Usually, these powders are highly charged when transported, loaded, dried, and mixed, due to repeated contact and separation between powder particles and the equipment. This results in the powders acquiring electrostatic charges. If not dissipated quickly, electrostatic charge buildup can cause inter-particle cohesion and adhesion to operating equipment, causing a production efficiency decrease. Incendiary electrostatic discharges such as spark discharges, cone discharges, and propagation brush discharges can also take place during industrial operations and may lead to fires and even explosions. Riedewald reported on an industrial fire in a pharmaceutical powder transfer operation [1]. A large number of studies involving electrostatic charge generation on powders, as well as electrostatic discharges and hazards during operation, have been reported. For example, Kaialy reviewed the main factors affecting the electrostatic charging of pharmaceuticals such as particle resistivity, size, shape, and surface roughness [2]. Gibson also reviewed industrial hazards related to electrostatic charges and discharges [3]. Our previous studies [4–6] investigated the electrostatic charging of polymer powder in mixing processes and the ignitability of aluminous, magnesium and polymer powders due to electrostatic discharges involving the effect of the ratio of N2 in the air on the ignitability.

In order to easily measure and confirm the electrostatic charge level of powders, a novel tribocharging system that utilizes the Faraday cup is presented, and the charging behavior of the sample powders is investigated in this paper.

Experimental Apparatus and method

Figure 1 shows the spiral air tribocharge apparatus presented in this study. The apparatus was divided into four main parts: the air supply part, the powder supply part, the tribocharge part, and powder collection part. The air supply part consisted of a compressor (Hitachi, PO3.7MA) that included an auto dryer (CKD, F3-217199), air controller, electronic valve, and switch. In this study, dried air was used because moisture in the air may have an effect on the electrostatic characteristics of the powder sample. The powder supply and tribocharge parts consisted of a powder port, a spiral air flow device, and a pipeline. The spiral air flow device (photo 1) was attached in front of the sample powder port in order to create spiral air inside the pipeline. The spiral air flow device had 8 holes that were 6 mm in diameter and that opened at a slant (60 degrees), which is very different from conventional models [7, 8, 9]. The straight pipeline was made up of stainless steel metal and was 1300 mm in length and 35 mm in diameter. In order to prevent external noise, the overall powder supply and tribocharge parts were shielded. The powder collection part was made up of a cyclone-type Faraday cup, unlike conventional models [7, 8, 9]. The cyclone-type Faraday cup had a separate metal container (30 mm in diameter) for easy measurement of the amount of collected powders. The other devices included two electrometers (Kethley,

6517A), a data logger (Graphtec, GL900), an electrostatic ionizer (Kasuga Denki Inc., KD-510W), an electrical balance (Shimazu, BL220H), and a timer.

The typical operating procedure for the tribocharge test is summarized as follows:
(1) Adjust the air pressure using the air pressure valve.
(2) Load 1 g of the sample powder into the powder port.
(3) Eliminate the initial charge of the sample powders using an electrostatic ionizer.
(4) Reset the electrometers (Q mode) connected to the pipeline and cyclone.
(5) Switch the air flow on, and run the sample powder through the pipeline where friction occurs.
(6) Measure the indication value when it stabilizes.
(7) Measure the amount of powder collected in the Faraday cup.
(8) Clean out the pipeline and cyclone part using an air spray in preparation for the next test.

The charge to mass ratio, q (nC/g), of the sample powder was obtained by dividing the charge, Q (nC), by mass, m (g), of the sample powder loaded into the Faraday cup:

$$q = \frac{Q}{m} \quad \cdots\cdots (1)$$

This experiment is mainly to determine how much the charge to mass ratio, q (nC/g), of the sample powders is affected by the size of the powder and the amount of air pressure. The cyclone-type Faraday cup and the pipeline were suspended electrically with insulators. All experiments were conducted in ambient conditions (air temperature: 22 ℃; relative humidity: 55 %).

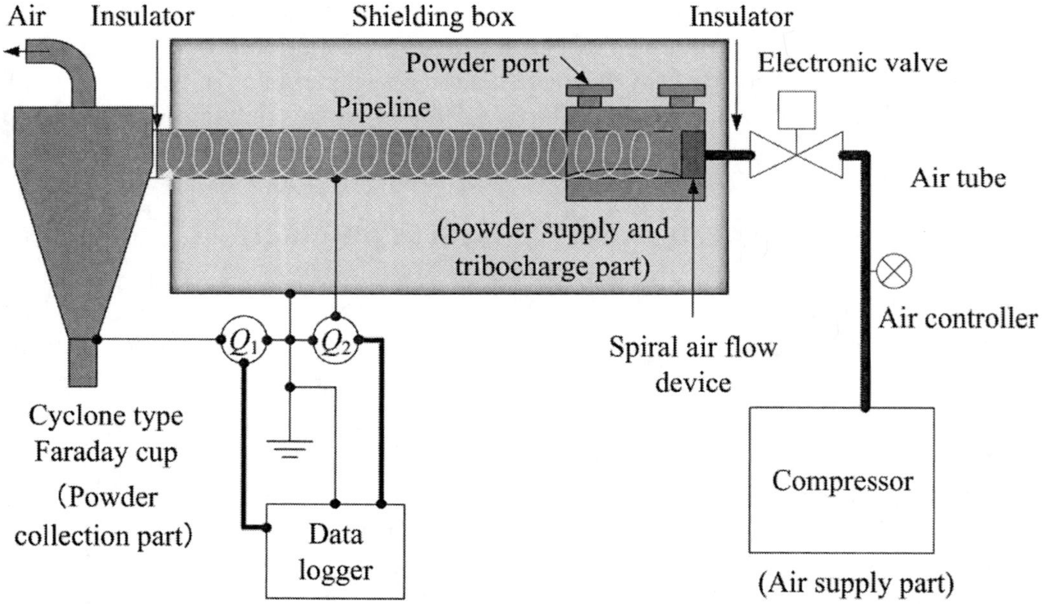

Fig. 1 Schematic of the spiral air tribocharge apparatus used in this study.

Photo 1 Structure of the powder port and spiral air flow device.

Sample powders

Glass beads are extremely good light reflectors and thus are widely used as the main component of the coating material for light-reflecting adhesive tape, background paint for traffic control sign boards, and paint for road signs. Four different grades of commercial glass beads were used in this experiment, as shown in Table 1. The particle size was determined using a laser diffraction system (LDA Win 1.21, dry type). Hereafter, they will be called by their abbreviation shown in Table 1. The analysis of the images from microscopy (Neoscope, JCM-500) revealed that the G-4 powder particles were spherical, as shown in photo 2. The other glass bead particles were the same shape as G-4.

Table 1. List of powder specimens used in experiments.

Specimen	Abbreviation	Diameter [μm]		
		D_{10}	D_{50}	D_{90}
Glass beads 1	G-1	22	62	85
Glass beads 2	G-2	93	131	175
Glass beads 3	G-3	130	180	241
Glass beads 4	G-4	162	225	294

D_X: X% volume average.

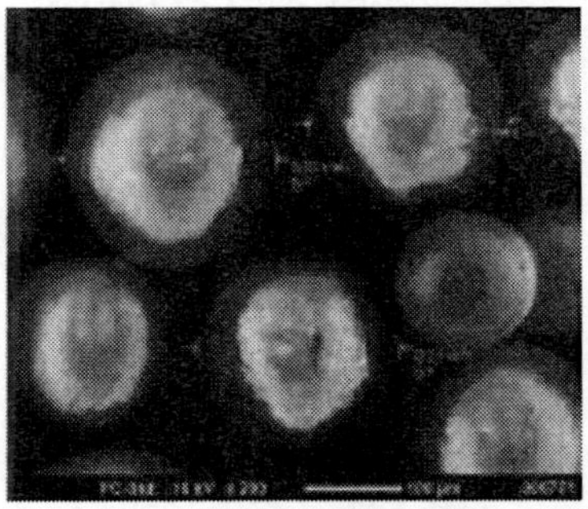

Photo 2 Shape of G-4 powder used in this experiment.

Experimental results and discussion

Figure 2 shows the typical change of charge detected over time when the G-3 particle was used and the air pressure was 0.15 MPa. It should be noted here that when no particle was used and only air was passed through the pipeline, the charge was detected as almost zero. The charge Q1 of the powder was found through the cyclone-type Faraday cup and then was confirmed through the charge Q2 obtained from the pipeline. As the powder particles passed through the pipeline for about 0.82 seconds, as seen in the Fig. 2, the charge clearly increased. Q1 and Q2 had almost the same trend and absolute values. This result confirmed the reliability of the test and showed that there was no disturbance by external noise.

In this study, all of the glass beads (G-1 to G-4) were negatively charged when stainless steel metal was used as the contact surface. The charge polarity of the materials relates to the work function (or ionization potential) of each material. The contact potential difference between two substances is defined by the difference between their fermi levels, which correlate with the work function. In general, materials with higher work functions tend to appropriate electrons from materials with lower work functions. This point will be carried out in our further study.

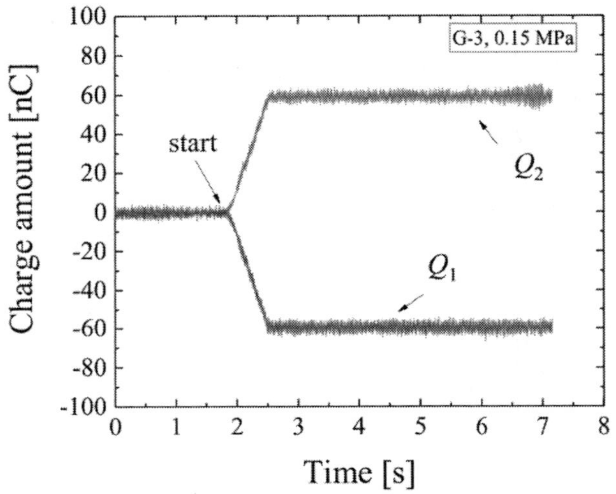

Fig. 2 Change of charge detected over time.

Figure 3 shows the charge to mass ratio of glass beads as a function of applied air. The range of applied air was between 0.1 MPa and 0.3 MPa in this study. When the applied air pressure was below 0.1 MPa, it was most likely not enough to send all of the powder through the pipeline. Each experiment was repeated five times, and the average values were reported. This graph does not contain an error bar, but these were about 10%. The results showed that the charge to mass ratio of both powders increased as the applied air increased. Although not shown in Figs. 2 and 3, the increase in air pressure also causes the slope of the graph to increase significantly. This can be explained by the fact that with increased pressure of applied air, the particles collide against the surface of the pipeline with more strength [9]. In order to investigate in more detail, the velocity of air at approximately 10 mm from the inside of the pipeline wall was measured using an anemometer (Kanomax, Anemomaster-6115). The approximate velocity of air flow was 9.5 m/s at 0.1 MPa, 12 m/s at 0.15 MPa, 15 m/s at 0.2 MPa, 20 m/s at 0.25 MPa, and 22 m/s at 0.3 MPa. It could be clearly seen that as the air pressure increased, the air velocity also increased. Another reason for this may be that because the air was applied in a spiral form, the particles may have collided more frequently against the pipeline when the air flow was increased. This exception will be investigated in detail in the future.

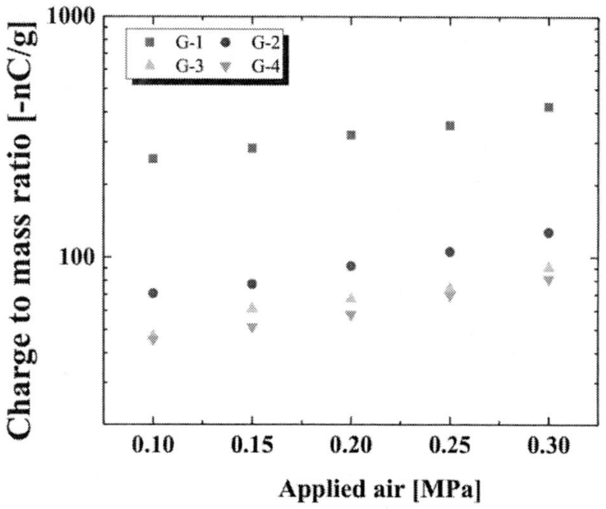

Fig. 3 Charge to mass ratio of glass beads as a function of applied air.

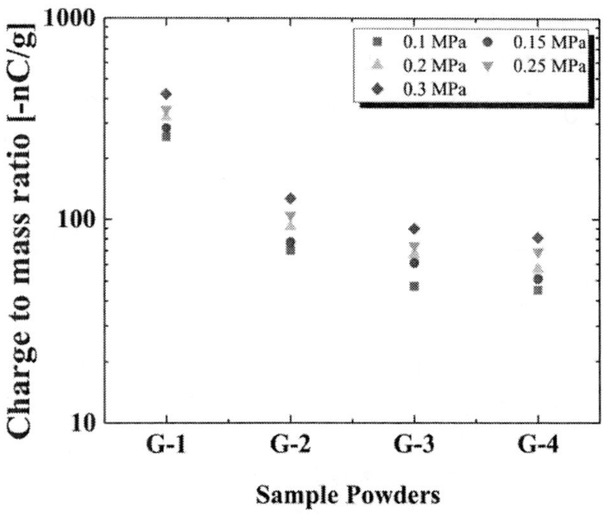

Fig. 4 Charge to mass ratio as a function of different glass bead sizes.

Figure 4 shows the charge to mass ratio as a function of different sizes of glass beads. The size distribution of each glass bead in this study was described in Table 1. The results showed that the charge to mass ratio of the glass beads increased as the powder size decreased. That is to say, smaller particles had a larger q than the larger ones. The total surface area of all of the smaller beads combined is greater than that of the larger ones, which results in a high charge to mass ratio development in the tribocharge. The value range of the q of the glass beads obtained in this study was about 40 nC/g to 100 nC/g. These values agreed with those obtained from the micronizing and pneumatic transfer process [10]. These values were high enough to generate incendiary electrostatic discharges during handling in industrial processes. This suggests that if glass beads are used with solvents in the cleaning process, they must be handled carefully, as incendiary discharges can ignite the solvents even though the glass beads themselves are nonflammable material.

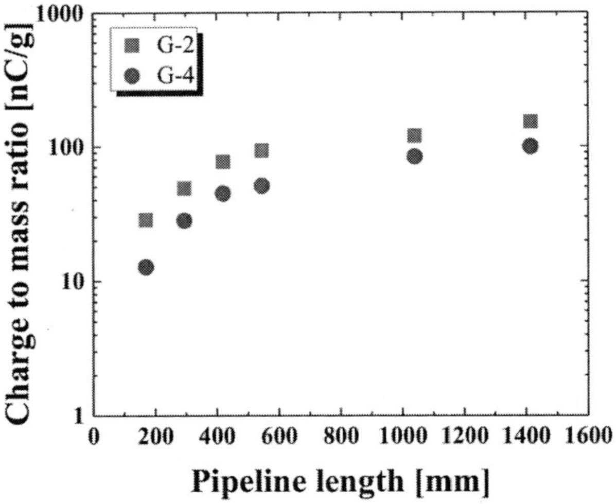

Fig. 5 Charge to mass ratio as a function of pipeline length.

Figure 5 shows the charge to mass ratio as a function of pipeline length. Six different pipeline lengths, *d*, were used: 170, 295, 420, 545, 1040, and 1415 mm. The glass beads used in this experiment were G-2 and G-4, and the air pressure was 0.35 MPa. Since the pipeline length changed with each test, it was difficult to attach the Faraday cup to the end of the pipeline. Therefore, in this study, only the value of q_2 was measured. As seen in Fig. 2, q_2 had the almost same value as q_1 but had an opposite sign. The length of the pipeline was measured starting from beginning of the powder port because the inside structure of the powder port was the same as that of the pipeline. Each experiment was repeated five times, and the average values were reported. The results showed that q_2 increased as the length of the pipeline increased for both sizes of glass beads. This can be explained by the fact that as the length increases, the time the glass beads come in contact with the pipeline also increases. The increase rate at the beginning section of the pipeline was especially large. It can also be seen through Fig. 5 that overall, G-2 had a larger q_2 than G-4 because of the total surface area. This confirms the reliability of the data shown in Fig. 4.

In addition, both beads seemed to reach saturation when the pipeline length was 1415 mm. It was compared to the maximum charge to mass ratio, q_c [nC/g] calculated from the well-known equation shown below [11]:

$$q_c = |\sigma_{max}|/(D \cdot \rho), \quad \text{--------------------------(2)}$$

where σ_{max} [kg/m^3] is the maximum surface charge density (27 μC/m^2); D [μm] is the particle diameter; and ρ is the particle density (1.5×10^3 kg/m^3).

In this study, the particle sizes of G-2 and G-4 were 131 μm and 225 μm, respectively, as shown in Table 1. The calculated maximum values q_c and experimental maximum values q_t are compared as shown in Table 2. The calculated and experimental values for both G-2 and G-4 agreed well with each other. Therefore, it can be seen that both G-2 and G-4 reached saturation.

Table 2. Calculated maximum values and experimental maximum values of the charge to mass ratio.

	q_c [nC/g]	q_t [nC/g]
G-2	137	151
G-4	80	99

Conclusion

In this paper, a novel tribocharging system that utilizes the Faraday cup was presented and the electrostatic charge behavior of glass beads was experimentally investigated. The results were as follows:
1) For all of the glass beads, the charge clearly increased as the glass bead particles passed through the pipeline. The charge amount of the glass beads and the charge amount obtained from the pipeline had almost the same trend and absolute values. This result confirmed the reliability of the test and showed that there was no disturbance by external noise.
2) All of the glass beads were negatively charged when stainless steel metal was used as the contact surface. The value range of the charge to mass ratio of the glass beads obtained in this study was about 40 nC/g to 100 nC/g.
3) The charge amount of the glass beads increased as the applied air increased. It could also be seen that as the air pressure increased, the air velocity also increased.
4) The charge amount of the glass beads increased as the powder size decreased due to the total surface area of the beads.
5) The charge amount of the glass beads (G-2, G-4) increased as the length of the pipeline. The increase rate at the beginning section of the pipeline was especially large.

Acknowledgments
The authors gratefully acknowledge Mr. Naoto Nogera for his assistance during the test.

References

1. F. Riedewald (2012): "A fire in a secondary pharmaceutical powder transfer operation," *Process safety progress*, 31, 4, pp. 390-392.
2. W. Kaialy W, "A review of factors affecting electrostatic charging of pharmaceuticals and adhesive mixtures for inhalation," *Int.J.Pharm.* Vol.503, pp.262-276.
3. N. Gibson (1997), "Static electricity — an industrial hazard under control," *J. Electrostatics,* Vol. 40-41, pp. 21-30
4. K. Choi, K. Choi, K. Nishimura (2015), "Experimental study on the influence of the nitrogen concentration in the air on the minimum ignition energies of combustible powders due to electrostatic discharges," *J. of Loss Prevention in the Process Industries.* Vol.34.pp.163-166.
5. K.S. Choi, H.Sakasai, and K. Nishimura (2015), "Experimental study on ignitability of pure aluminum powders due to electrostatic discharges and Nitrogen's effect," *J. of Loss Prevention in the Process Industries.* Vol.35.pp.232-235.

6. K.S. Choi, H. Sakasai, and K. Nishimura (2015), "Minimum ignition energies of pure magnesium powders due to electrostatic discharges and Nitrogen's effect," *J. of Loss Prevention in the Process Industries.* Vol.41.pp.144-146.
7. N. Masui, and Y. Murata (1981), "charging tendency of the flowing powder through a metal tube (II), " *The five Annual meeting of Institute of Electrostatic Japan,* pp. 244-247.
8. S. Watano, S. Saito, and T. Suzuki (2002), "Feedback control of powder electrostatic charge in a pneumatic conveying process," *J. Soc. Powder,Japan*, Vol. 39, pp. 496-502.
9. P. Boschung, M. Glor (1980), "Methods for investigating the electrostatic behavior of powders, J. of elelctrostatic," Vol. 8, pp. 205-219.
10. IEC50404(2004), "Electrostatics- Code of practice for the avoidance of hazards due to static electricity," p. 66.
11. M.Glor (1988), "Electrostatic hazards in powder handling," *Research studies press LTD.*, p.26.

509266 Spray-Drying of a Layered Double Hydroxide Nanosuspension

Abstract:

Over the past few years, layered double hydroxide (LDH) has attracted particular interests in both academia and industry. LDH is 2D anionic clay of brucite-like layered structure with various organic or inorganic anions located between layers. The diverse composition of LDH, and its associated physical and chemical properties open up new possibilities in various applications. LDHs are successfully applied as precursors in synthesis of supported catalysts, sorbents, carriers delivering bioactive molecules, matrices containing active pharmaceutical ingredients for control release, electrochemical biosensors, etc. LDHs are mainly synthesized in a form of nanosuspensions. The application of nano-sized sorbents or catalysts in the liquid-phase processes requires a special technique to separate them from the final product. An uncontrollable agglomeration of solid nanoparticles makes their industrial application impossible in a dry phase. Moreover, the toxic effects of nanoparticles on human health and ecosystem have been discussed in the literature.

Utilization of larger micron-size agglomerates of nanoparticles is one of the alternative ways to safely handle nanoparticles in the dry phase on the industrial scale. Spray drying is one of the well-known techniques that have been widely applied for production of microspheres from nano-sized particles. Using this technique, it is possible to control the size distributions as well as the morphology of spray-dried agglomerates. However, there are only a limited number of publications related to spray drying of LDH nanosuspensions. Moreover, spray-dried LDH agglomerates of rather wide size distribution, low surface area and irregular shape were reported in the literature.

In this study, we explore the effect of process parameters, such as spraying air flow rate, drying air temperature, slurry concentration and slurry feed rate on the size distribution, morphology, specific surface area and pore volume of LDH agglomerates fabricated by spray drying of nanosuspensions.

The round-shaped agglomerates with a diameter of more than 2 and less than 15 micrometers were obtained by spray drying of suspensions of nanoparticles with median diameter of 171 nm without addition of dispersant and binder. Our results confirmed that the spraying air flow rate has the dominant effect on the agglomerate median size and the width of agglomerate size distribution as well as on the agglomerate specific surface area. The small agglomerates with narrow size distribution were collected during drying at the high flow rate of spraying air. Our results also indicate that lowering the temperature of drying air and using the dilute nanosuspensions promote generation of small agglomerates.

The specific surface area of spray-dried LDH agglomerates is in the range of 49 to 101 m^2/g. The agglomerates of high specific surface area were obtained at low flow rate of spraying air, high slurry feed rate of concentrated slurry and using low temperature drying air. These results also confirm that drying at low drying rate yields to formation of agglomerates of high surface area. Similar trends were observed in terms of the variation of total pore volume with drying process parameters.

Thus, by adjusting the spray drying conditions, it is possible to produce agglomerates of LDH nanoparticles satisfying the product requirements.

Copyright © American Institute of Chemical Engineers. All rights reserved.

513829 Impact of Spheroidization of UO_2 Powders on the Filling of Press Molds

Abstract:

In industry, the use of powders is very common. This underlines the interest in understanding the flow of powders in process lines (hopper, silos, transportâ€¦) and finding the link between powder physical characteristics and its rheological properties. This study deals with the flowing properties of a UO2 powder in a powder metallurgy process for nuclear fuel fabrication. A reproducible filling of the press molds is required to avoid mechanical grinding of the sintered pellets.

A powder is characterized by its particle size distribution but also by the morphology of particles. Morphology has a major impact on the cohesion and then on the flowability of powders [1]. UO2 powders are generally made up agglomerates which are composed of aggregates, themselves formed by crystallites (Figure 1). Agglomerates are primarily governing the powder flowing behavior. Their characteristics sometimes do not allow them to meet the expectations of the process, especially for the filling of the press molds.

For this reason, in a powder manufacturing process, a spheroidization operation (Figure 2) could precede the shaping of pellets to improve the filling properties.

In this study we observe the influence of particles morphology on the rheological properties of UO2 powders. Whereas there is no standard and universal method to characterize the flow of powders [2], there is direct (shear cells) and indirect methods (dynamic flowability measurements like permeability). We use both methods: the indirect methods based on qualitative measures of powder cohesion, the direct methods to determine the powder cohesion for a specific consolidation state through the Freeman shear cell [3]. While a shear cell measurement permits to determine the powder flow in quasi-static system, a permeability measurement shows the flow index in dynamic system. We identify the link between theses measurements and the quality of the filling.

References:
[1] Xiaowei Fu and others, «Effect of Particle Shape and Size on Flow Properties of Lactose Powders», Particuology, Advances in Characterization and Modeling of Particulate Processes, Apr. 2012.
[2] P. Juliano, B. Muhunthan, and G. V. Barbosa-Cánovas, « Flow and shear descriptors of preconsolidated food powders », J. Food Eng., vol. 72, no 2, p. 157-166, Jan. 2006.
[3] J. Schwedes and D. Schulze, « Measurement of flow properties of bulk solids », Powder Technol., vol. 61, no 1, p. 59-68, Apr. 1990.

Copyright © American Institute of Chemical Engineers. All rights reserved.

508203 Modelling a Twin Screw Granulator Using the Discrete Element Method

Abstract:

Wet granulation is a process used to create larger stable agglomerates (granules) from fine powders. This has many desirable outcomes such as improving flowability, compactibility, and homogeneity. Granulation is commonly employed in the food, pharmaceutical, detergent, and fertilizer industries, but despite its wide adoption it is often inefficiently operated, with high recycle ratios in continuous processes and high rejection rates in batch processes. Wet granulation is the most common type of granulation and, in pharmaceutical applications, it is a critical step in tablet manufacturing that affects the uniformity and compactibility of the final dosage form.

Traditionally batch granulation was the favoured granulation process method in the pharmaceutical industry due to the challenges faced from adopting a continuous approach, such as cost or inability to monitor product quality reliably. Since the introduction of Quality by Design (QbD) and Process Analytical Technology (PAT) by the FDA in 2003 there has been a move towards new manufacturing techniques such as continuous processing.

Although tremendous efforts have been made to gain scientific insight into the granulating process, a fundamental understanding of wet granulation is still lacking due to the complexity of the mechanisms involved. With twin screw granulation becoming a popularly employed method of wet granulation, an in-depth understanding of particle enlargement in the granulating process is necessary in order to improve the quality of the final product without the need for large-scale Design-of-Experiment studies.

Despite the extensive experimental research carried out for twin screw granulators in recent years, there has been little computational work carried out. This paper employs the Discrete Element Method (DEM) to study a 25 mm diameter, GEA ConsiGmaâ„¢ 1 twin screw granulator with a typical 60Â° forward configuration for kneading elements. The DEM simulations were conducted using the commercial code EDEM with a DEM contact model developed for cohesive solids implemented through an API. The contact model is based on an elasto-plastic contact with adhesion and uses hysteretic loading and unloading paths to model the elastic-plastic contact deformation. In these simulations, the adhesion is used to capture the effect of the binder liquid without the complication of modelling the liquid directly. The adhesion parameter is a function of the plastic contact overlap. The model has previously been shown to be able to predict the stress-history-dependent behaviour depicted by a flow function of the material. In this study the effect of particle size, gap tolerance and cohesion on the residence time distribution and local solid fraction at different elements in the granulator is assessed. The level of cohesion used is related back to uniaxial test results which can be used to measure the flowability of a cohesive granular solid.

Copyright © American Institute of Chemical Engineers. All rights reserved.

508216 Hybrid Multiscale Modelling of a Twin Screw Granulator

Abstract:

Wet granulation is a process used to create larger stable agglomerates (granules) from fine powders. This has many desirable outcomes such as improving flowability, compactibility, and homogeneity. Granulation is commonly employed in the food, pharmaceutical, detergent, and fertilizer industries, but despite its wide adoption it is often inefficiently operated, with high recycle ratios in continuous processes and high rejection rates in batch processes. Wet granulation is the most common type of granulation and, in pharmaceutical applications, it is a critical step in tablet manufacturing that affects the uniformity and compactibility of the final dosage form.

Modelling processes such as wet granulation is not straightforward, but may be achieved using techniques such as Population Balance Modelling (PBM) and Discrete Element Method (DEM). PBM groups particles into different classes based on their properties and uses empirically derived differential equations to describe key processes such as aggregation or breakage. However, although PBM tracks the size and number of particles over time, it cannot account for the distribution of the particles and the effect of this local variability.

By contrast, DEM simulations track individual particles, and as such, phenomena such as segregation or liquid migration can be studied. DEM simulations also consider the effect that the geometry and its dynamics have on the particles' behaviour, which is something that PBM cannot account for directly. However, DEM simulations of fine materials can be computationally expensive and this often results in reduced numbers of particles being used to make the simulation computationally tractable. This is usually achieved through increased, unrealistic particle sizes or the simulation of isolated local zones without considering adjacent zones which may affect the behaviour.

Traditionally, large-scale Design-of-Experiment (DoE) studies would be needed to estimate key rate parameters for use in PBM modelling of granulation and this can be both costly and time-consuming. DEM simulations provide an alternative method to estimate these parameters. By coupling the two simulation methods, one can utilise the key benefits of each method to better simulate complex processes.

This paper presents a hybrid approach to modelling twin screw granulation via DEM–PBM coupled simulations to study a 25 mm diameter, GEA ConsiGma™ 1 twin screw granulator with a typical 60° forward configuration for kneading elements. The DEM simulation was conducted using the commercial code EDEM with a DEM contact model developed for cohesive solids implemented through an API. The PBM simulation is carried out with the commercial code gFORMULATE which is built on top on the advanced modelling platform gPROMS. A bi-directional coupling is implemented with information shared between the two simulation tools at specified intervals. Preliminary results for this coupled approach are presented and discussed.

Copyright © American Institute of Chemical Engineers. All rights reserved.

509577 Blocking Rules for Discharging Granular Materials from a Flat Bottom Hopper

Abstract:

Hoppers and silos are widely used in storing granular materials in various industries, such as agricultural, chemical, food and pharmaceutical industries. Three key aspects need to be considered when hoppers are used to store powders and grains: 1) stress analysis; 2) hopper blockage ; 3) mass flow rate. A hopper must be designed strong enough to sustain the stress generated from the bulk materials during storage and discharge. On the other hand, a hopper must be designed properly so that the stored bulk materials can be discharged, i.e. without blockage and formation of ratholes and arches. Furthermore, the mass flow rate needs to be predicted and controlled. In this study, a combined experimental and numerical study was performed to determine the blocking rule for discharging granular materials from a flat bottom hopper. Granular materials of various sizes were considered and the minimal orifice size through which the bulk materials can be discharged was determined experimentally using a Flodex tester and numerically using DEM. The dependency of the minimal

orifice size on particle size was explored. An semi-empirical model was also developed. Using this model, the critical orifice size at which the flow will stop, i.e. blockage takes place, can be determined from the mass flow rate data. It was found that excellent agreement between the numerical results and experimental measurements was obtained. Both experimental and numerical results showed that, even for spherical particles, layered dead zones could be formed during the discharge from the flat bottom hopper. In addition, mono-layered dead zones were developed when large particles were used. Furthermore, the developed blocking rules were also verified experimentally and numerically.

Copyright © American Institute of Chemical Engineers. All rights reserved.

508310 Characterization of Mesoscopic Structure in Cohesive Powder, Neat or Blended, By X-Ray Computed Tomography and Prediction By the Discrete Element Method

Abstract:

Understanding and predicting structural characteristics of cohesive powder systems is a technical challenge in many powder technology processes and applications. The discrete element method (DEM) [1] is a promising approach to model the structure and flow behavior of cohesive granular systems. Advances in experimental techniques such as X-ray computed tomography (XRCT) have now enabled characterization of fine cohesive powder systems with sub-micron resolution. Combined with standard bulk powder characterization methods (e.g. density), this technique allows for mesoscopic structural characteristics of cohesive powder systems to be compared with DEM simulation. In this study, DEM and XRCT are used to characterize the mesoscale structure, under controlled consolidation conditions, of two moderately cohesive 'model' (roughly spherical) powders with distinct material properties. The DEM simulations are based on a constitutive elasto-plastic model [2] that makes use of the JKR theory [3] to account for cohesive interparticle forces. Density and void structure obtained from DEM simulation are compared to characterization results obtained from XRCT image analysis. XRCT and DEM simulation approach is applied to examine structures of binary mixtures of the two model powders. The ability of DEM simulation and the contact model to accurately predict structural characteristics of single component and binary granular systems is discussed.

References

[1] P. A. Cundall and Otto D.L. Strack. A discrete numerical model for granular assemblies. Geotechnique, 29(1):47–65, 1979.

[2] C. Thornton and Z. Ning. A theoretical model for the stick/bounce behavior of adhesive, elastic-plastic spheres. Powder technology, 99(2):154– 162, 1998.

[3] K.L. Johnson, K. Kendall, and A.D. Roberts. Surface energy and the contact of elastic solids. In Proceedings of the Royal Society of London A: Mathematical, Physical and Engineering Sciences, volume 324, pages 301–313. The Royal Society, 1971.

Copyright © American Institute of Chemical Engineers. All rights reserved.

511638 Numerical Research of Hydrodynamics in Gas-Solid Micro Fluidized Beds

Abstract:

Due to the advantages of less energy consumption, more safety and good operability, the miniaturization of fluidized bed, or named micro fluidized bed, has been extensively studied by a lot of researchers. The Micro Fluidized Bed Reaction Analyzer (MFBRA) was adopted by the Institute of Process Engineering (IPE), Chinese Academy of Sciences (CAS), to study the gas-solid reaction kinetics under isothermal differential condition[1, 2]. In the MFBRA, the hydrodynamics of micro fluidized beds play a vital role on the heat and mass transfer and which is a key to the accurate measurement of gas-solid reaction kinetics. In order to optimize and design the MFBRA, numerous experiments have been conducted to study the hydrodynamics of micro fluidized beds[3, 4]. Unfortunately, those works mostly focused on the bed pressure drop and minimum fluidization velocity, nevertheless, the detail information, such as local solid volume fraction and particle velocity profile in micro fluidized beds are reported rarely. It is because the existing experimental techniques, such as, optical fiber and capacitance probes are hard to measure those local parameters without disturbing the flow field.

The objective of this work is to study the hydrodynamics of micro fluidized beds and to investigate the isothermal characteristics during kinetics testing of graphite powder combustion in MFBRA. In this study, Discrete Particle Model (DPM) and Two-Fluid Model (TFM) are selected to investigate micro fluidized beds, respectively. The DPM is carried out by the commercial software EDEM-Fluent Coupling simulations. And for TFM, we used the Eulerian multiphase model in the ANSYS Fluent. From our numerical simulations, the detailed hydrodynamic characteristics of micro fluidized beds are explored systematically by comparing different column sizes and superficial gas velocity. Meanwhile, it is found that both DPM and TFM have the ability to model the MFBRA under practical operation condition with high superficial gas velocity. In the term of isothermal characteristics in micro fluidized beds, the temperature field distribution as well as the heat transfer between heat-carrier and graphite powder are explored by the combination of simulations and experiments.

Reference:

1. Xu, G., *Micro fluidized bed reaction kinetics analyzer. Supported by instru-ment development project of Chinese Academy of Sciences, China, Project numberY2005014. (in Chinese).* 2005.
2. Yu, J., et al., *Isothermal differential characteristics of gasâ€"solid reaction in micro-fluidized bed reactor.* Fuel, 2013. **103**: p. 29-36.
3. Liu, X., G. Xu, and S. Gao, *Micro fluidized beds: Wall effect and operability.* Chemical Engineering Journal, 2008. **137**(2): p. 302-307.
4. Guo, Q.j., Y. Xu, and X. Yue, *Fluidization Characteristics in Micro-Fluidized Beds of Various Inner Diameters.* Chemical Engineering & Technology, 2009. **32**(12): p. 1992-1999.

Copyright © American Institute of Chemical Engineers. All rights reserved.

507312 An Improved Bubble-Based Drag ModelÂ for Accurate Coarse-Grid Two-Fluid Modeling of Geldart a Powder Bubbling Fluidization

Abstract:

An improved bubble-based energy minimization multi-scale (EMMS) drag model was formulated in which the original bubble-based EMMS drag model at macroscale bed level was extended to microscale sub-grid level for accurate coarse-grid two-fluid modeling (TFM) of Geldart A powder bubbling fluidization. At both the macroscale and microscale levels, the physically heterogeneous structures were modeled by three pseudo phases, i.e., bubble phase, emulsion phase, and inter-phase, where conservation and stability condition equations were applied. Through such scale and phase decompositions, the improved bubble-based EMMS drag model was formulated through a two-step macro-to-micro scheme, where the obtained drag coefficient is related to phase volume fractions, velocities, and accelerations within each coarse grid. Modeling results show that TFM combined with the improved bubble-based EMMS drag model can accurately reproduce the heterogeneous structures for a laboratory-scale Geldart A powder bubbling fluidized bed. At relatively coarse grid, the predicted axial and radial distributions of solid concentration agree very well with experimental measurements. This indicates that our proposed improved bubble-based EMMS drag model incorporated TFM is promising for accurate and economical simulation of engineering-scale Geldart A powder bubbling fluidization.

Copyright © American Institute of Chemical Engineers. All rights reserved.

507860 Numerical Simulation on Fine Particle Transport Behaviour in Electrostatic Precipitators

Abstract:

Electrostatic precipitators (ESPs) are a major air pollution control device to remove particles from treated gas. Extensive research has been done on several aspects such as electrostatics, fluid dynamics, charging mechanism, and particle dynamics. The overall particle-collection efficiency of a modern industrial wire-plate ESP exceeds 99%. However, the removal efficiency of submicron particles by conventional ESPs is as low as 70â€"80% owing to the low charge carried by fine particles and the increase in mobility with decreasing size. Submicron particles possess a high superficial area, thus favouring toxic heavy metals and other pollutants. Therefore, fine particles are collected with much lower efficiency and further research is required to improve this aspect.

In this work, a numerical study on the charging and transport behaviour of fine particles has been carried out based on wire-plate electrostatic precipitators (ESPs) with multiple wire electrodes. The effect of the applied wire voltage, inlet height, and wire distance on particle charging and transport behaviour, and the influence of the precipitator structure on particle trapping are analysed in detail. The 2-D numerical model involves electric field, space-charge density, gas flow, and particle trajectories. The fluid-flow velocity is higher than 1 m/s in this study and the EHD flow has a negligible. A Lagrangian approach for particle transport is utilized with the trajectories of individual particles tracked by solving their equation of motion. The electric field and gas flow are calculated using the finite-volume method. The flow simulation is done using the commercial software FLUENT with other parts coded as user-defined functions. The following main results can be drawn:

(1) It was observed that the Y-displacement of the particles exhibited a universal increase for all particle sizes, especially for larger particles, at low injection positions ($Y < 25$ mm). The larger the particle size, the higher the Y-displacement of particles because larger particle sizes improve both the particle charging level and electric field strength. No significant improvements were found for submicron particles.

(2) Increasing the spacing of the discharge electrode within a certain range and therefore increasing the electric field, it was observed that the particle-trapping effect improved and the outlet level exhibited a universal increase for all particle sizes, especially for large particles.

(3) The ESP models M1 and M2 were used to investigate their influence on the outlet height. Except for particles with a diameter of 2 μm, all other particle sizes showed an increased outlet height in M2 compared to that in M0. In addition, the different discharge-electrode arrangements, especially in M2, exhibited significant effects on the charging and transport behaviour of particles with a diameter of 1 μm.

(4) Particles with an inlet height of 5 and 10 mm were studied in M1 and M2 precipitators. The changed discharge-electrode arrangement improves particle trapping, especially in M2. This is because of the electron trajectories being closer to the discharge electrodes (M2), giving the particles more charge and leading to a greater electric field.

Copyright © American Institute of Chemical Engineers. All rights reserved.

507879 Lattice Boltzmann Simulations of Porous Particulate Flows

Abstract:

The comprehensive understanding of the particle-fluid interaction is very important for fluid-particles system, and thus many studies have been conducted for the particle-fluid system[1-3] of solid impermeable particles. In the real industrial fields such as the catalyst process in chemical engineering[4], the particles usually possess the porous structure which will have effects on the fluid-particles interaction[5-6]. However, there has been little open literature discussing the particle-fluid system of porous permeable particles. Therefore, the objective of the present study is to investigate the effect of the porosity and permeability on the porous particulate flows which are formulated by the volume-averaged governing equations[7] in terms of intrinsic phase average velocity. The lattice Boltzmann equation model (LBE)[8] is used to solve the governing equations. Firstly, we validate the LBE model by the numerical results existing in the literature for the flow around and through one porous particle. Then, the porous particles settling against gravity in a quiescent fluid, i.e., the DKT process[9] for porous particles, is investigated. The effects of the porosity and Darcy number on the porous particulate flows are studied numerically in details. Also, we present the simulation results of the fluidization of 512 porous particles. Our results demonstrate the significance of the porous structure of particles for the particles dynamics behaviors.

Reference

[1] Ladd, A.J.C., 1994. Numerical simulations of particulate suspensions via a discretized Boltzmann equation Part I. Theoretical foundation. J. Fluid Mech., 271, 285-310.

[2] Hill, R.J., Koch, D.L., Ladd, A.J.C., 2001. The first effects of fluid inertia on flows in ordered and random arrays of spheres. J. Fluid Mech. 448, 213-241.

[3] Van der Hoef, M.A., Beetstra, R., Kuipers, J.A.M., 2005. Lattice Boltzmann simulations of low-Reynolds-number flow past mono- and bidisperse arrays of spheres: results for the permeability and drag force. J. Fluid Mech. 528, 233-254.

[4] Zhao, Y.F., Li, H., Ye, M., Liu, Z.M., 2013. 3D numerical simulation of a large scale MTO fluidized bed reactor. Ind. Eng. Chem. Res. 52, 11354-11364.

[5] Noymer, P.D., Glicksman, L.R., Devendran, A., 1998. Drag on a permeable cylinder in steady flow at moderate Reynolds numbers. Chem. Eng. Sci. 53, 2859-2869.

[6] Zhu, Q.Y., Chen, Y.Q., Yu, H.Z., 2014. Numerical simulation of the flow around and through a hygroscopic porous circular cylinder. Comput. Fluids 92, 188-198.

[7] Wang, L., Wang, L.-P., Guo, Z.L., Mi, J.C., 2015. Volume-averaged macroscopic equation for fluid flow in moving porous media. Int. J. Heat Mass Trans. 82, 357-368.

[8] Chen, S.Y., Doolen, G.D., 1998. Lattice Boltzmann method for fluid flows. Annu. Rev. Fluid Mech. 30, 329-364.

[9] Fortes, A.F., Joseph, D.D., Lundgren, T.S., 1987. Nonlinear mechanics of fluidization of beds of spherical particles. J. Fluid Mech. 177, 467-483.

Copyright © American Institute of Chemical Engineers. All rights reserved.

508370 Modeling of a Novel Multi-Particle Collision Model for Gas-Solid FlowsÂ

Abstract:

Fluidized beds are widely applied in the chemical, petrochemical, metallurgical, environmental and energy industries due to their favourable mass and heat transfer characteristics. In past few decades computational fluid dynamics has proven to predict hydrodynamics of fluidized beds. DPM (Euler-Lagrange) [1] and TFM (Euler-Euler) [2] are two most common modelling approaches for gas-solid flow in fluidized beds. In TFM the solid phase itself is treated as a continuum phase interpenetrating with the gas phase. Since the concept of individual particles has disappeared, the effects of particle-particle interactions in the TFM has to be incorporated in a statistically averaged way, via the solid phase pressure and viscosity. TFM is preferred model to study large scale fluidized beds but model suffer from high numerical diffusion. And the KTGF theory in TFM assume elastic collisions and is limited only to spherical particles [2]. In DPM, the gas phase is considered as the continuous phase whereas the particle trajectories are obtained by integrating Newtonâ€™s equations of motion. Particle-particle and particle-wall interactions are explicitly taken into account using various physical models such as the soft sphere model. DPM proves to be accurate modelling of gas-particle flow. However, lagrangian models quickly become computationally too expensive because of the large number of particle interactions. Therefore coarse graining of discrete particles is an emerging model that can handle large number of particles effectively. The new coarse-grained model developments are of vital importance for the advancement of large-scale modelling of gas-particle flows that are flexible to extended for polydispersity, and non-spherical particles. In this study a novel lagrangian multi-particle collision (MPC) method is developed for predicting dense particulate-fluid flow.

The MPC solver is developed in OpenFoam [5], where the fluid phase is treated as a continuum and the solids are modelled by lagrangian particles. The MPC model grew out of the multiphase particle-in-cell (MP-PIC) model [3], that employs a fixed Eulerian grid, and particles are traced in a lagrangian fashion, including a parcel approach where each simulated particle represents a large collection of real particles. Each parcel follows Newtonâ€™s equations of motion, where inter-phase momentum transfer is accounted through mapping and interpolating quantities back and forth between parcel locations and the eulerian grid. Particle collisions in MP-PIC are not fully resolved but are derived from particles stresses that are theoretically calculated from Harris and Crighton [4] stress closure. In the MPC model, we implemented a functional form for the particle stress tensor where the coefficients are obtained computationally by means of DPM simulations. The idea is to extend the model from spherical particles to non-spherical particles, for which theoretical particles closures are currently not available in the literature. The viscous stress and random forces are typically ignored in MP-PIC model due to their insignificant effect [3]. Importantly, we incorporated viscous stress and random forces representing scattering of particles by collisions, which are obtained from DPM simulations by measuring the actual variance. Details of model development and validation is presented in this study.

Results from MPC model are compared with MPPICFoam and, with DPM simulations from DEM-LIGGGHT [6], and MFiX [7] code respectively. Pseuduo-2D fluidized beds of dimension 0.01x0.15x0.45 m is considered for comparison with experiments [8]. Comparison is made on the basic of pressure drop, void fraction, bubble dynamics and solids motion. We observed that MPC is in good agreement with DPM, and experiments, when compared to MPPICFoam. Unrealistic over-packing of solids is predicted in MPPIC due to limitation of Harris and Crighton [4] model and the velocity field updates, whereas in MPC these limitations overcomes and no over packing of particles is predicted. Solid circulations are in good agreement with DPM results, where solid moving above in the center and downward near to the wall. Time-average porosity plots revels the path of bubble motion and is in close agreement with results of DPM. In conclusion a novel multi-particle collision model is developed and validated to study gas-solid fluidized beds. The model is efficient to handle large number of particles. In future study, we plan to extend the model for dense flow of non-spherical particles.

Acknowledgments

The authors thank the European Research Council for its financial support under its consolidator grant scheme, contract no. 615906 (NonSphereFlow).

References

1. N.A. Patankar, D.D. Joseph, 2001. Modeling and numerical simulation of particulate flows by the Eulerian Lagrangian approach, International Journal of Multiphase Flow, Volume 27, Issue 10, October 2001, Pages 1659-1684.
2. Gidaspow, D., 1994. Multiphase flow and fluidization continuum and kinetic theory description, Academic Press, Boston.
3. Andrews, M.J., O'Rourke, P.J., 1996. The multiphase particle-in-cell (MP-PIC) method for dense particulate ows, Int. J. Multiphase Flow 22, 379-402.
4. Harris, S.E., Crighton, D.G., 1988. Solutions, solitary waves and voidage disturbances in gas-fluidized beds, J. Fluid Mech. 195, 437.
5. OpenFOAM, 2015. The Open Source CFD Toolbox, Ver. 3.0.0. ESI Group, Paris, France.
6. https://www.cfdem.com/liggghtsr-open-source-discrete-element-method-particle-simulation-code
7. https://mfix.netl.doe.gov/mfix/
8. M. Goldschmidt, 2001. PhD thesis, Hydrodynamic Modelling of Fluidised Bed Spray Granulation.

Copyright © American Institute of Chemical Engineers. All rights reserved.

508490 Dpm Analysis of Large Fluidized Catalytic Cracking (FCC) Reactors

Abstract:

DPM Analysis of Large Fluidized Catalytic Cracking (FCC) Reactors

Abstract:

Computational Fluid Dynamic (CFD) modeling is a powerful tool for equipment design and development and is widely used at UOP, particularly for the reactor and regenerator design of FCC process unit. Through CFD modeling, a technical solution was developed for a large capacity FCC process unit purposing an elevated amount of propylene production.

A key challenge in FCC process design is how to optimize the size of the reactor vessel, which is a function of the number and size of reactor cyclones (catalyst separationdisengagers). UOP developed a new cyclone cluster configuration via CFD study of the reactor internals, including UOP's VSS™ riser termination device and reactor cyclones, which reduce the reactor size significantly compare to the conventional design, while maintaining effective catalyst / vapor separation and mechanical reliability. The objective of the current CFD study is to evaluate the catalyst separation efficiency and performance of the new riser termination device technology, which accommodate more cyclones in a smaller reactor vessel.

To better understand the flow pattern, flow distribution and the gas residence time inside the UOP VSS™ , the analysis was broken down into two sections. Part 1 focused on single gas phase only and in part 2 the particles behavior is studied using the Discrete Phase Model (DPM) in Fluent.

The single gas flow analysis results showed that the flow in the VSS chamber section has the same rotational pattern as in conventional FCC/VSS units, where the catalyst will separate from the gas products. Also the gas flow distribution at each chamber arm is uniform, which is critical to catalyst separation efficiency.

The DPM analysis confirmed an equal catalyst particle distribution through all the cyclones and a high efficiency separation in the riser termination device. Having this information from CFD allows for the new design to move forward.

Copyright © American Institute of Chemical Engineers. All rights reserved.

509623 Catalytic Reactor Design Using Multiphase CFD

Abstract:

Catalytic Reactor Design Usng Multiphase Cfddimitri Gidaspow the 2011 Book , "Computational Gas-Solids Flows and Reacting Systems ", Edited By S. Pannala, M. Syamlal and Tom O'Brien ,Engineering Science Reference, Shows the Progress We Have Made in Using Multiphase CFD in Modeling Fluidized Bed Chemical Reactors and Other Multiphase Systems, Such As the Pyroplastic Flow Emitted from Volcanoes. in This Presentation the Design of Two Types of Catalytic Reactors Will be Described: A Fischer_ Tropsch Slurry Bubble Column Reactora Polymerization Reactor the CFD Design of a New Slurry Bubble Column Reactor Was Already Described By Gidaspow ,Et Al in Chemical Engineering Science 134,784-799 (2015).in This Presentation the Design Will be Extended to Obtain 100, 000 Barrels of Diesel per Day, Twice That of the Largest Commercial Reactors. the Production of Synthesis Gas Using Oxygen or Carbon Dioxide in Catalytic Fluidized Beds Needed for the Fischer-Tropsch Reactor Will Also be Discussed Briefly. for the Polymerization Reactor Our CFD Code Described in " Computational Techniques " By D. Gidaspow and V. Jiradilok , Nova Science, 2009 Was Modified for Polymerization Reactors. Such Reactors and Their CFD Modeling Are Described in a Chapter in Pannala's Et Al Book from Professor Rodney Fox's Group. They Point out That the Fine Particle (80 microns) Catalyst Radial Distribution, Is a Remaining Issue the Computations to be Presented Will Show That the Catalyst Will Have a High Concentration at the Reactor Walls for Various Operating Conditions. This Leads to a High Rate of Reaction at the Reactor Walls and May Result in Melting of the Polymers and Sheet Formation , Forcing Reactor Shut-Down. in Some of Our Simulations There Is Large Bubble and Slug Formation Similar to Those Mentioned in Fox's Chapter. Acknowledgement : I Thank Dr. Mayank Kashyap of SABIC for a Valuable Discussion and His Papers " Application of Multiphase Flow CFD in the Gas Phase Polymerization Process " Fluidization XV. May 24,2016 and a Related Paper Presented at Doe NETL, July 2016

Copyright © American Institute of Chemical Engineers. All rights reserved.

509627 Simulation of Large Particle Turbulent Fluidization in Riser Reactors By Coarse Grain DEM-CFD

Abstract:

The complex two-phase flow in the riser is an important element of many industrial circulating fluidized bed reactors. In some cases, e.g. gas-solid polymerization, relatively large particles are transported upward in a turbulent regime, in which the high gas velocity results mostly from heat removal requirements. Strong solids acceleration, radial and axial velocity profiles for both the gas and solid, intense clustering and inhomogeneities make it impossible to reliably predict the riser head or local bed density profile. Detailed numerical simulations proved capable of showing some of the peculiarities of the flow, but limitations in the scale, lack of grid independence, need for special drag corrections and sub-grid models still hinder confident use of such tools in design and scale-up, particularly in industrial applications characterized by particle size and Reynolds numbers that are not well covered by the available scientific literature.

In the present contribution, coupled Discrete Element Method-Computational Fluid Dynamics (DEM-CFD) simulations are carried out using a coarse graining technique for the solid particles. According to this method, small number of particles are grouped into a computational particle (*parcel*) and represented by it as a spherical "lump," to reduce the computational cost of the simulations and allow lab-scale systems to be represented in full detail. The investigation focuses on the influence of the coars-graining degree, i.e. the number of real particles represented by a *parcel*, on the simulated flow hydrodynamics both in terms of its advantages (a quantification of the computational savings) and disadvantage, primarily the real capability of the method to resolve the behaviour observed and produce results in line with experimental findings. The gas and solid motion are studied, with the aim to characterize the clustering phenomena observed during the up-flow and its effect on the pressure drop. Assuming prescribed gas and solid inlet velocity fields, the potential tendency to show core-annulus flow, the intensity of solids clustering and the local bed density within the riser are examined at different number of particles per *parcel*.

Copyright © American Institute of Chemical Engineers. All rights reserved.

509640 Numerical Study of Particles Shape Effects on Solid-Liquid Fluidizations

Abstract:

Numerical study of particles shape effects on solid-liquid fluidizations

Esmaeil Abbaszadeh Molaei, Zongyan Zhou*

Laboratory for Simulation and Modelling of Particulate Systems, Department of Chemical Engineering, Monash University, Clayton, VIC 3800, Australia

Solid-liquid fluidizations have been observed in many industries due to their high heat and mass transfer rate, high operation flexibilities, and reduced back mixing of fluid and solid phases. In most of these multiphase operations, particle properties such as size, density, and shape may experience significant changes during the process because of attrition, coalescence, comminution or chemical reactions. The difference in particle properties can result in significant segregation problem in liquid fluidized beds. To gain an understanding of solid-liquid fluidisation behaviour, most existing studies, either experimentally or numerically, have focused on solid flow behaviour with mono-sized particles or binary and ternary mixtures with different particle sizes or densities. In spite of substantial studies on the hydrodynamics of liquid fluidized beds, the effects of particles shape on the solid-liquid flow behaviour are still less reported. Except for two physical experiments, the previous studies considered particles as spheres, and few studies have been remarked on liquid fluidization of non-spherical particles. However, particles are generally irregular in most of the processes such as ore beneficiations using liquid-solid fluidized beds separator. Particle shape can cause different flow behaviour in liquid fluidizations rather than that observed for ideal spherical particles. Nonetheless, the answers to some fundamental questions such as how and why particle shape affects flow phenomena are not clear, and hence still poorly understood.

Therefore, in this work, a combination of computational fluid dynamics (CFD) and discrete element model (DEM) [CFD-DEM] is used to examine the effects of particle shape on the flow characteristics in liquid fluidizations. In the simulations, ellipsoidal particles are used as they can represent a wide range of particle shapes from oblate to prolate particles. Different aspect ratios are employed, and results are analysed mainly in terms of particle flow patterns, pressure drop, and bed expansion. In addition, force and particle orientation analysis is also conducted to examine the effect of particle shape and explain the occurrence of entrainment phenomenon.

The results show that ellipsoidal particles intend to enter to freeboard region and entrainment phenomenon occurs, and consequently, the interface between the bed and freeboard cannot be observed clearly. In addition, the pressure gradient variation with bed height is uniform for spheres; however, this uniformity is replaced with â€œdistorted parentheses-shapeâ€ profile for disc-shape and elongated particles. Finally, bed expansion analyses show that spherical particles have the lowest bed expansion. On the other hand, disc-shape particles intend to expand more significantly than spherical and elongated particles. This causes the entrainment phenomenon which has been found to be due to the effects of particles shape on particles terminal velocity.

* Corresponding author. Tel: 61 3 9905 0846; email: zongyan.zhou@monash.edu.

Copyright © American Institute of Chemical Engineers. All rights reserved.

514318 CFD Investigation on Gas-Solids Flow and Heat Transfer in Two Fluidized Catalyst Cooler

Abstract:

Our previously experimental work showed that the internal circulation of solids in an annular catalyst cooler (ACC) had an ability to enhance the heat transfer between bed and wall, comparing with the base catalyst cooler (BCC). The current work simulated in detailed hydrodynamics of gas-solids flow and heat transfer to advance our understanding of the ACC by using the two-fluid model. The effective particle diameter d_pâŽ was determined to 170 Î¼m and the specularity coefficient Ï† was 0.3 in the 3D simulations. The catalyst cooler can be divided to four distinct regions including draft tube region, annulus region, bottom region and particle diffluence region according to the distributions of solids volume fraction and solids velocity. The effect of gas-solids flow on bed-to-wall heat transfer was compared in the different regions of the BCC and ACC. And the effects of operational gas velocity and bed structure were discussed. The results showed that the proper selection of the ratio of gas velocities and the position of the heat transfer tube were needed to form a stable internal solids circulation in the ACC. The ACC had a combined hydrodynamic feature of up-flow and down-flow catalyst coolers with bigger solids volume fraction and smaller particle resident time, which are beneficial for improving the heat transfer between solids and wall.

Copyright © American Institute of Chemical Engineers. All rights reserved.

Hydrodynamics and mixing characteristics of a new-type particle mixer

Mengxi Liu, Chunxi Lu, Zhenliang Meng

State Key Laboratory of Heavy Oil Processing, China University of Petroleum, Beijing 102249, PR China

E-mail: mengxiliu@sina.com (M. Liu)

Abstract:

In order to further enhance the particle mixing in the internally circulating fluidized bed, especially along the radial direction, a new-type particle mixer with slots on the draft tube was proposed in the present work. The hydrodynamics of the mixer was investigated by using a multi-scale computational fluid dynamics with the EMMS drag. The prediction in terms of cross-sectionally averaged solid holdup and solid velocity were compared to experimental data for validation. The results show that in draft tube a low pressure region can be observed near the slots and in annulus the pressure in the vicinity of the slots seems larger than that in the region between the adjacent slots, which induces horizontal particle immigration and mixing. Nearly 25.4% to 29.7% of gas in the draft tube bypasses and flows through the slots into the annulus. The ratio of the mass flux of the tracer gas bypassing into the annulus to that of total tracer gas is approximately equal to 0. With increase of the superficial gas velocity in the draft tube from 0.3 m/s to 0.45 m/s, the total solid mass flux increases from 4.87 kg/s to 10.86 kg/s, while the mass flux through slots decreases from 2.69 kg/s to 0.52 kg/s.

1. Introduction

The characteristics of particle flow and mixing have a considerable influence on the mass and heat transfer, as well as the product quality. According to the previous study, the mixing degree of particles along the axial and radial direction is quite different [1-5]. Mostoufi and Chaouki [4] discovered that the diffusivities varied from 3.3×10^{-3} m^2/s to 5.6×10^{-2} m^2/s along the axial direction and 2.6×10^{-4} m^2/s to 1.5×10^{-3} m^2/s along the radial direction. Berruti et al. [6] also revealed that the particle mixing rate for radial direction was an order of magnitude smaller than for the axial direction. Therefore, the weak particle flow along the radial direction greatly limits the mixing ability of a fluidized bed. Internally circulating fluidized bed (ICFB) has been widely recognized as an efficient fluidized bed reactor, in which orderly and macroscopic particle circulation occurs. Compared with the conventional fluidized bed, ICFB offers multiple particle circulation between the draft tube and the annulus, which leads to relatively strong particle convection, diffusion and shear, and thereby improves the particle mixing performance along the radial direction [7]. In order to further enhance the particle mixing for some processes in which a uniform particle mixing is required, a new-type particle mixer with slots on the draft tube was proposed in the present work.

With increase of the computational power, the computational fluid dynamics (CFD) provides an additional tool to predict particle flow and mixing in the local region, which is difficult with the experimental methods. In particular, the gas-solid flow in a gas-solid fluidized bed can be accurately described by the energy minimization multi-scale (EMMS) model [8]. In this paper, the hydrodynamics of the mixer was investigated by using a multi-scale CFD based on the EMMS drag model [9, 10].

2. Experimental apparatus, Materials and Operating conditions

The structure of the particle mixer is shown in Fig.1. The mixer has a height of 5 m and an inner diameter of 286 mm. A draft tube with a height of 1 m and an inner diameter of 215 mm is coaxially mounted into the column. The four rectangular slots of 80×40 mm are symmetrically opened on the draft tube at a height of h=860 mm, in order to strengthen the particle radial flow. A gas distributor plate and a ring distributor are installed below the draft tube and the annulus respectively. The height of the gap between the gas distributor plate and the draft tube is 60 mm. For the sake of the convenience, the mixer is divided into six zones: the pre-mixing zone, feed influence zone, draft tube zone, gas-solid separation zone, annulus zone and the bottom zone. A perforated plate is mounted at the bottom of the pre-mixing zone, which is connected with a central downcomer, extending into the draft tube. Particles are first mixed in the pre-mixing zone and then the pre-mixed particles flow downward through the central downcomer and into the feed influence zone. In the feed influence zone, the pre-mixed particles are mixed with the particles moving upward in the draft tube, flow upward and circulate between the draft tube and the annulus. In the annulus, particles move downward to the bottom. Then a part of particles flow out of the mixer and the other part flow horizontally into the draft tube and recirculate.

The solid particles used in the experiment were FCC catalyst (ρ_p=1498 kg/m^3), with an average diameter of 76 μm. Air (ρ_g=1.225 kg/m^3) was introduced into the mixer to fluidize the particles. The superficial gas velocity ranged from 0.3 to 0.45 m/s in the draft tube (based on the cross-sectional area of the draft tube) and remained at 0.05m/s in the annulus (based on the cross-sectional area of the annulus). The mass flux of cold particles remained at 13.8 kg/m^{2}s. As we only investigated the influence of slots on the flow and mixing of particles in this paper, no hot particles were induced into the mixer.

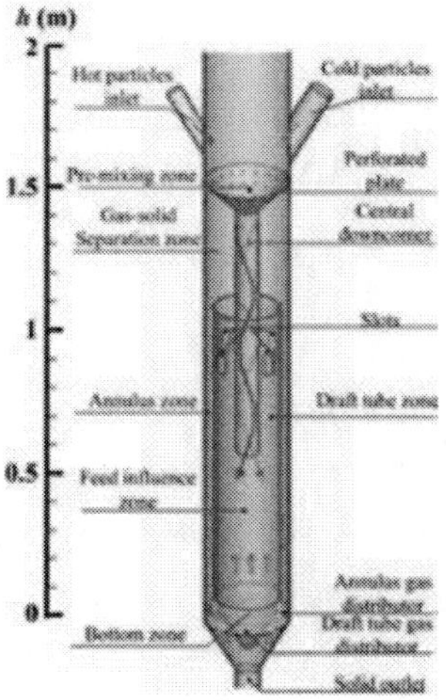

Fig. 1 Schematic diagram of the new-type particle mixer

3. CFD model and simulation method

The hydrodynamics of the mixer was performed using the Eulerian- Eulerian of Ansys Fluent

®13.0. The governing equations [11] include a set of mass and momentum equations for each phase, as shown in Table 1.

A 3-D mesh containing 580,000 cells was generated by using Gambit®2.4. The no-slip boundary was assigned for the gas phase and the partial-slip boundary condition developed by Johnson and Jackson [12] was used for the solid phase. In order to ensure a steady-state operation of the mixer, simulations were first performed for 30 s, and then the time-averaged data was analyzed for another period of 10 s. Other parameters settings in the model are shown in Table 2.

Table 1 Governing equations

Gas-phase Continuity equation	$\frac{\partial}{\partial t}(\varepsilon_g\rho_g)+\nabla\cdot(\varepsilon_g\rho_g\boldsymbol{u}_g)=0$
Solid-phase Continuity equation	$\frac{\partial}{\partial t}(\varepsilon_s\rho_s)+\nabla\cdot(\varepsilon_s\rho_s\boldsymbol{u}_s)=0$
Gas-phase momentum equation	$\frac{\partial}{\partial t}(\varepsilon_g\rho_g\boldsymbol{u}_g)+\nabla\cdot(\varepsilon_g\rho_g\boldsymbol{u}_g\boldsymbol{u}_g)=-\varepsilon_g\nabla p+\nabla\cdot\boldsymbol{\tau}_g+\varepsilon_g\rho_g\boldsymbol{g}-\beta(\boldsymbol{u}_g-\boldsymbol{u}_s)$
Solid-phase momentum equation	$\frac{\partial}{\partial t}(\varepsilon_s\rho_s\boldsymbol{u}_s)+\nabla\cdot(\varepsilon_s\rho_s\boldsymbol{u}_s\boldsymbol{u}_s)=-\varepsilon_s\nabla p-\nabla p_s+\nabla\cdot\boldsymbol{\tau}_s+\varepsilon_s\rho_s\boldsymbol{g}+\beta(\boldsymbol{u}_g-\boldsymbol{u}_s)$
Granular temperature equation	$\frac{3}{2}[\frac{\partial}{\partial t}(\varepsilon_s\rho_s\Theta_s)+\nabla\cdot(\varepsilon_s\rho_s\Theta_s\boldsymbol{u}_s)]=(-p_s\boldsymbol{I}+\boldsymbol{\tau}_s):\nabla\boldsymbol{u}_s+\nabla\cdot(\kappa_s\nabla\Theta_s)-\gamma_s-3\beta\Theta_s$
Gas-phase stress	$\boldsymbol{\tau}_g=\varepsilon_g\mu_g[\nabla\boldsymbol{u}_g+(\nabla\boldsymbol{u}_g)^T]-\frac{2}{3}\varepsilon_g\mu_g(\nabla\cdot\boldsymbol{u}_g)\boldsymbol{I}$
Solid-phase stress	$\boldsymbol{\tau}_s=\varepsilon_s\mu_s[\nabla\boldsymbol{u}_s+(\nabla\boldsymbol{u}_s)^T]+\varepsilon_s(\lambda_s-\frac{2}{3}\mu_s)(\nabla\cdot\boldsymbol{u}_k)\boldsymbol{I}$
Solid-phase pressure	$p_s=\varepsilon_s\rho_s\Theta_s[1+2(1+e_s)\varepsilon_s g_0]$
Solid-phase shear viscosity	$\mu_s=\mu_{s,col}+\mu_{s,kin}+\mu_{s,fr}$
Where	$\mu_{s,col}=\frac{4}{5}\varepsilon_s\rho_s d_p g_0(1+e_s)\sqrt{\Theta_s/\pi}$, $\mu_{s,kin}=\frac{10\rho_s d_p\sqrt{\Theta_s/\pi}}{96\varepsilon_s(1+e_s)g_0}[1+\frac{4}{5}g_0\varepsilon_s(1+e_s)]^2$, $\mu_{s,fr}=\frac{p_s\sin\phi}{2\sqrt{I_{2D}}}$
Radial distribution function	$g_0=[1-(\varepsilon_s/\varepsilon_{s,max})^{1/3}]^{-1}$
Diffusion coefficient	$\kappa_s=\frac{150\rho_s d_p\sqrt{\Theta_s/\pi}}{384(1+e_s)g_0}[1+\frac{6}{5}g_0\varepsilon_s(1+e_s)]^2+2\rho_s\varepsilon_s^2 d_p(1+e_s)g_0\sqrt{\Theta_s/\pi}$
Collisional energy dissipation	$\gamma_s=\frac{12(1-e_s^2)\varepsilon_s^2\rho_s g_0\Theta_s^{3/2}}{d_p\sqrt{\pi}}$
Heterogeneity index	$H_d=\frac{\beta_e}{\beta}$, $\beta_e=\beta H_d$
Drag based homogeneous structure	$\beta=\begin{cases}\frac{3}{4}C_d\frac{\varepsilon_s\varepsilon_g\rho_g\lvert\boldsymbol{u}_g-\boldsymbol{u}_s\rvert}{d_p}\varepsilon_g^{-2.65}, & \varepsilon_g>0.8\\ 150\frac{\varepsilon_s^2}{\varepsilon_g}\frac{\mu_g}{d_p^2}+1.75\varepsilon_s\frac{\rho_p}{d_p}\lvert\boldsymbol{u}_g-\boldsymbol{u}_s\rvert, & \varepsilon_g\le 0.8\end{cases}$
Where	$C_d=\begin{cases}0.44 & Re_s\ge 1000\\ \frac{24}{Re_s}(1+0.15Re_s^{0.687}), & Re_s<1000\end{cases}$, $Re_s=\frac{\varepsilon_g\rho_g d_p\lvert\boldsymbol{u}_g-\boldsymbol{u}_s\rvert}{\mu_g}$

Table 2 Simulation Parameters

Gas-solid model	Eulerian-Eulerian
Solids Pressure	Lun et al.
Particle-wall specularity coefficient	0.5
Gas-wall friction coefficient	No-slip
Pressure-Velocity Coupling	Simple
Gradient discretization	Least squares cell based
Momentum discretization	Second-order upwind
Volume fraction discretization	Quick
Max. number of iterations per time step	40
Pressure outlet (atm)	1
Inventory (kg)	56
Incipient solid holdup	0.6

4. Results and discussion

4.1 Model validation

Fig. 2 gives the distribution of cross-sectionally averaged solid holdup and particle velocity in the mixer. As shown in Fig. 2(a), the solid holdup is larger in the bottom zone (h=0.04 m) and due to the accelerated upward motion of particles, the solid holdup decreases with increase of the axial height (when $h \leq 0.16$ m). The cross-sectionally averaged solid holdup is nearly constant in the draft tube and gas-solid separation zone of $0.4 \leq h \leq 1.25$ m. The cross-sectionally averaged particle velocity is shown in Fig. 2(b). The particle velocity is smaller in the bottom zone and keeps almost invariable in the draft tube and the gas-solid separation zone. Compared with the experimental data, the particle velocity seems slightly higher, while the solid holdup in the simulation agrees reasonably with the experimental results.

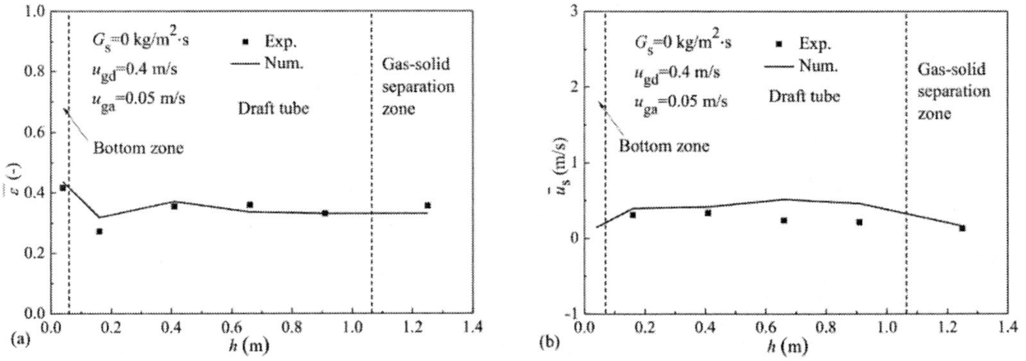

Fig. 2 Distribution of cross-sectionally averaged solid holdup and particle velocity as a function of axial heights

4.2 Pressure distribution near the slots

Fig. 3 shows the time-averaged contours of static pressure on the cross-section of h=0.9 m, the cross-section on the middle of the slots. It is seen that in the draft tube a low pressure region can be observed near the slots, smaller than the pressure in the region between the adjacent slots. This pressure drop forced particles inside the draft tube to move towards the slots and pass through them, which significantly facilitates the mixing of the particles along the radial direction. In the annulus the pressure in the vicinity of the slots seems larger than that in the region between the adjacent slots, inducing horizontal particle immigration and mixing.

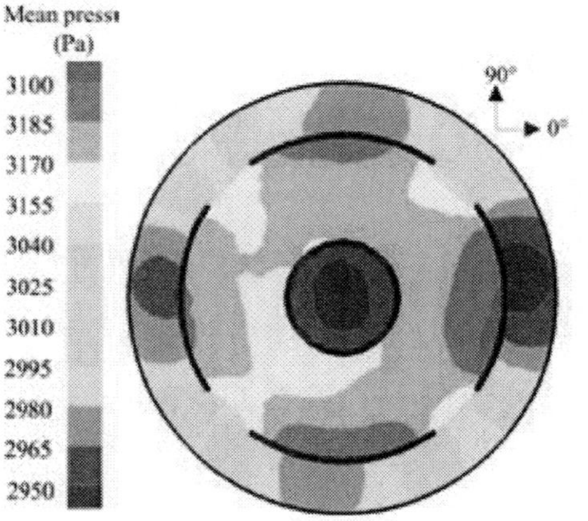

Fig. 3 Time-averaged contours of static pressure at the cross-section of the slots (u_{gd} =0.4 m/s, G_s=13.8 kg/m^{2}s, h=0.9 m)

4.3 Gas bypassing

In the present work, the gas bypassing behavior was investigated by CFD simulation, and a tracer gas having identical physical properties with air was defined. After the mixer running stably, the tracer gas was introduced into the mixer from the gas distributor plate to replace the original gas. A cross section (h=0.06 m) on the exit of the annulus was selected and the mass flux of the tracer gas through the cross section was registered. The results are shown in Fig. 4. f_B refers to the ratio of the mass flux of the tracer gas through the cross section (h=0.06 m) to that of total tracer gas. It is seen that f_B varies between -0.05 and 0.05 with a frequency of 10 Hz, indicating an unstable gas bypassing from the draft tube to annulus. This suggests that the gas bypassing into the annulus was finally brought out of the annulus by the circulating particles. The average f_B is approximately equal to 0, much smaller than that in a conventional ICFB (usually within 15%) [13], because massive particle circulation in the mixer makes it difficult to bypass from the draft tube to annulus.

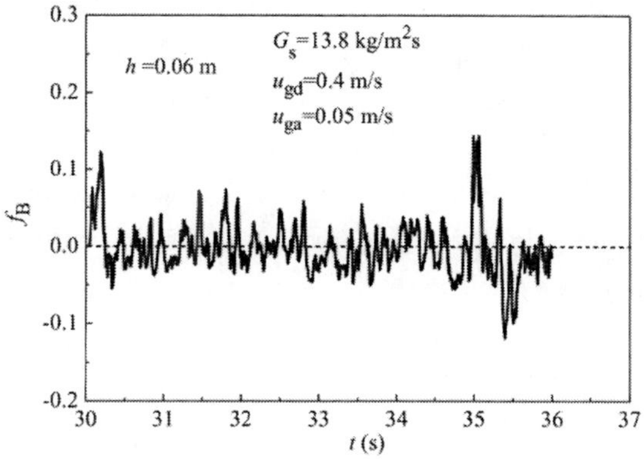

Fig. 4 Gas bypassing from draft tube to annulus in the bottom zone

As discussed previously, in the mixer there exists a pressure gradient from the draft tube to the vicinity of the slots, and then the annulus. Such a pressure difference leads to horizontal flows of particles and gas from the draft tube to the slots and then to the annulus. In the mixer the upside and downside of the slots is located at h=0.94 m and 0.86 m, respectively. Mass flux of the gas bypassing through the slots was obtained by monitoring the gas flux through the lower cross section (h=0.7 m) and the upper cross section (h=1 m) in the draft tube. Fig. 5 shows the gas mass flux under different operating conditions. The total gas mass flux in the draft tube increases with increasing gas velocity in the draft tube. With increase of the gas velocity from 0.3 to 0.45 m/s, the mass flux of the gas bypassing through the slots increases from 0.157 to 0.221 kg/m^{2}s. But the mass flux of bypassing gas does not increase linearly with the gas velocity. It is also seen that the mass ratio of the gas through the slots to that in the draft tube drops from 29.7% to 25.4%.

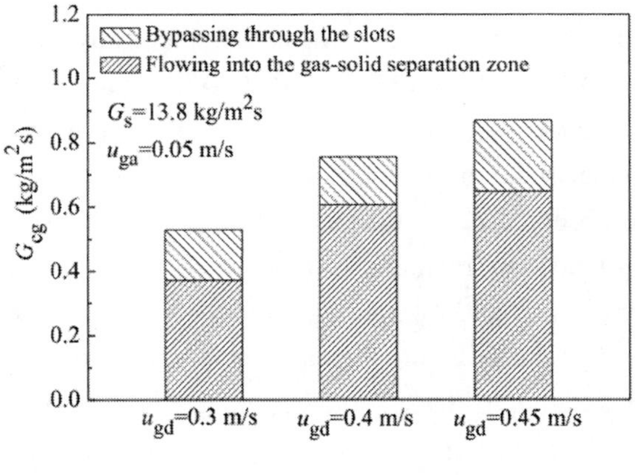

Fig. 5 Gas mass flux through the slots and the gas-solid separation zone as a function of superficial gas velocity in the draft tube

4.4 Particle mass flux

Fig. 6 shows the particle mass flux in the mixer. It is seen that the mass flux increases with increase of the superficial gas velocity in the draft tube. The mass flux through the slots accounts for about 50% of the total mass flux when u_{gd}=0.3 m/s. Because the axial particle velocity rose with increase of the superficial gas velocity from 0.3 m/s to 0.45 m/s in the draft tube, making it difficult to change the direction of the particles. Meanwhile, the radial particle velocity above the draft tube increases. Consequently, with the increase of the superficial gas velocity in the draft tube , the mass flux through the slots considerably decreased, from about 55.3% of the total mass flux (about 2.69 kg/s) at u_{gd}=0.3 m/s to only 4.8% (about 0.52 kg/s) at u_{gd}=0.45 m/s. Meanwhile, the total particle mass flux in the mixer increased from 4.87 to 10.86 kg/s as u_{gd} increased from 0.3 to 0.45 m/s.

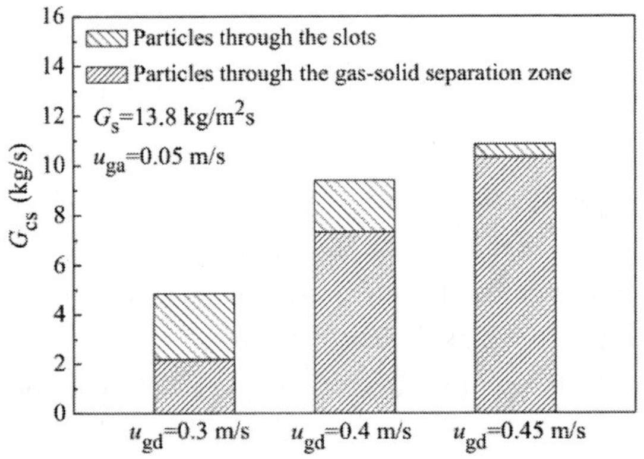

Fig. 6 Particle mass flux through the slots and the gas-solid separation zone as a function of superficial gas velocity in the draft tube

5. Conclusions

The characteristics of particle flow and mixing have a considerable influence on the mass and heat transfer, as well as the product quality. In order to further enhance the particle mixing in the ICFB, especially along the radial direction, a new-type particle mixer with slots on the draft tube was proposed in the present work. The hydrodynamics of the mixer was investigated by using a multi-scale computational fluid dynamics with the EMMS drag. The prediction in terms of cross-sectionally averaged solid holdup and solid velocity were compared to experimental data for validation. The results show that in draft tube a low pressure region can be observed near the slots and in annulus the pressure in the vicinity of the slots seems larger than that in the region between the adjacent slots, which induces horizontal particle immigration and mixing. Nearly 25.4% to 29.7% of gas in the draft tube bypasses and flows through the slots into the annulus. The ratio of the mass flux of the tracer gas bypassing into the annulus to that of total tracer gas is approximately equal to 0. With increasing the superficial gas velocity in the draft tube from 0.3 m/s to 0.45 m/s, the total solid mass flux increases from 4.87 kg/s to 10.86 kg/s, while the mass flux through slots decreases from 2.69 kg/s to 0.52 kg/s.

Notation

G_s	solid mass flux (kg/m^2 s)	G_{cg}	gas mass flux in the mixer (kg/m^2 s)
G_{cs}	solid mass flux in the mixer (kg/s)	h	height (m)

p	pressure (Pa)	R	radius of the reactor (mm)
r	radial position of measurement points (mm)	t	time (s)
u_{ga}	superficial gas velocity in the annulus (m/s)	u_{gd}	superficial gas velocity in the draft tube (m/s)
$\overline{u}$	cross-sectionally averaged axial particle velocity (m/s)		

Greek

ε_g	voidage	ε_s	solid holdup
$\overline{\varepsilon_s}$	cross-sectionally averaged solid holdup	ρ	density (kg/m^3)

References

[1] F. Niklasson, H. Thunman, F. Johnsson, B. Leckner, Estimation of solids mixing in a fluidized-bed combustor, Ind Eng Chem Res, 41 (2002) 4663-4673.

[2] L.H. Shen, M.Y. Zhang, Effect of particle size on solids mixing in bubbling fluidized beds, Powder Technol, 97 (1998) 170-177.

[3] L.H. Shen, M.Y. Zhang, Y.Q. Xu, Solids Mixing in Fluidized-Beds, Powder Technol, 84 (1995) 207-212.

[4] N. Mostoufi, J. Chaouki, Local solid mixing in gas-solid fluidized beds, Powder Technol, 114 (2001) 23-31.

[5] D.Y. Liu, X.P. Chen, Lateral solids dispersion coefficient in large-scale fluidized beds, Combust Flame, 157 (2010) 2116-2124.

[6] F. Berruti, D.S. Scott, E. Rhodes, Measuring and Modeling Lateral Solid Mixing in a 3-Dimensional Batch Gas Solid Fluidized-Bed Reactor, Can J Chem Eng, 64 (1986) 48-56.

[7] M.M. Fang, K. Luo, S.L. Yang, K. Zhang, J.R. Fan, Computational Fluid Dynamics-Discrete Element Method Investigation of Solid Mixing Characteristics in an Internally Circulating Fluidized Bed, Ind Eng Chem Res, 52 (2013) 7556-7568.

[8] J. Li, M. Kwauk, Particle-fluid Two-phase Flow: the Energy-minimization Multi-scale Method, Metallurgical Industry Press, Beijing, 1994.

[9] Z.S. Shi, W. Wang, J.H. Li, A bubble-based EMMS model for gas-solid bubbling fluidization, Chem Eng Sci, 66 (2011) 5541-5555.

[10] K. Hong, Z.S. Shi, W. Wang, J.H. Li, A structure-dependent multi-fluid model (SFM) for heterogeneous gas-solid flow, Chem Eng Sci, 99 (2013) 191-202.

[11] D. Gidaspow, Multiphase flow and fluidization: Continuum and Kinetic Theory Descriptions, Academic Press, Boston, 1994.

[12] P.C. Johnson, R. Jackson, Frictional–collisional constitutive relations for granular materials, with application to plane shearing, J. Fluid Mech. 176 (1987) 67–93.

[13] B.H. Song, Y.T. Kim, S.D. Kim, Circulation of solids and gas bypassing in an internally circulating fluidized bed with a draft tube, Chem Eng J, 68 (1997) 115-122.

514498 Power Spectral Density Analysis of Pressure Signal in 18 m Circulation Fluidized Bed Riser

Abstract:

Studies on the gas-solids flow characteristics using Power Spectral Density based on the pressure data from different axial locations in a riser of diameter 80 mm and height 18 m. Results show that the pressure signal fluctuation is mainly affected by gas-solids interaction, which included gas-solids friction, particle-particle collision and particle-particle aggregation. With the increasing solids circulating rate, frequency of particle interactions including particle-particle collision and particle-particle arrogation become strong and the pressure fluctuation also increased. It can be seen that the pressure signal fluctuation intensity decreased gradually with the increasing axial position of the riser. The low frequency but high power phenomenon nearly disappears in the fully developed zone near the top of the riser. Based on the results, it can be concluded that the gas-solids interaction is stronger at the bottom of riser. On the other hand, the gas-solid interaction is more homogeneous, and the pressure fluctuation is relatively weak in the fully developed zone.

Copyright © American Institute of Chemical Engineers. All rights reserved.

514514 Characteristics of Pressure Fluctuations in Particle-Transport Inclined Pipe of a Circulating Fluidized Bed

Abstract:

In a large circulating fluidized bed device, the characteristics of pressure fluctuation in a particle-transport inclined pipe under different solid circulating flux were investigated based on measuring the dynamic pressure of gas-solid two-phase flow by using FCC equilibrium catalyst as the particles. The experimental results showed that with increase of particle mass flux, the flow patterns of FCC particles above and below the butterfly valve in the inclined pipe both successively changed, and the fluctuation amplitude of dynamic pressure increased. The standard deviation of the dynamic pressure below the butterfly valve possessed a linear relation with the particle mass flux, which could be used to characterize the real-time dynamic pressure in the inclined pipe. By wavelet analysis, it is indicated that there exists a main frequency caused by inclined pipe flow structures in the pressure fluctuations above and below the butterfly valve, besides, there is a secondary frequency in the pressure fluctuation below the butterfly valve caused by gas turbulence and gas-solid interactions. With the increase of particle mass flux, the wavelet energy ratio of the secondary frequency gradually decreased. The fluctuation pressure caused by discharge in the inclined pipe is a low frequency pressure and the excitation source of inclined pipe vibration.

Copyright © American Institute of Chemical Engineers. All rights reserved.

508323 Dynamic Modeling of Attrition and Reactions in Circulating Fluidized Bed Reactors

Abstract:

Fluidized bed reactors are widely applied in chemical processes due to the excellent gas-solid contact. A problem that occurs in fluidized beds is the mechanical stress that particles are exposed to during the fluidization, which causes particle attrition. Additionally, fluidized beds are often operated in batch mode, which will cause ever-changing operating conditions with time. This makes fluidized bed modeling especially challenging, because the distributed particle properties, like size, wetness or reaction state, have to be tracked over the whole course of the simulation.

At the Institute of Solids Process Engineering at Hamburg University of Technology, novel process simulation software named DYSSOL is developed, particularly tailored for solids processes. The software is based on the flowsheeting concept. This means that chemical unit operations, like for example chemical reactors or separators, are put into process modules, which are interconnected into a process flowsheet and then mathematically solved. To make dynamic simulations available, new dynamic models had to be proposed for the involved equipment. The models themselves are based on empirical and semi-empirical correlations for hydrodynamics and kinetic particle reaction models are integrated.

The dynamic flowsheeting software was then used to simulate a system of interconnected fluidized bed reactors, which are used for Chemical Looping Combustion of solid fuels. This process is especially interesting, because heterogeneous reactions take place. Further, the process is a reactor-regenerator system causing the reactive solid to be oxidized and reduced and, hence changing its conversion state. Attrition plays a major role for the cost effective operation of the process.

In a small lab scale bubbling bed reactor, the jet and bubbling attrition rate constants for the solid oxygen carrier were experimentally determined under ambient conditions. In a small stand-alone cyclone the cyclone attrition was investigated. Furthermore, the reaction kinetics of the solid material with fuel gases was experimentally determined. All this information was included into the flowsheet simulation of a pilot scale circulating fluidized bed reactor, which is actually operated in our institute to deliver validation data.

The dynamic flowsheet simulation of the pilot plant, which was enhanced with kinetic and attrition information of smaller scale facilities, was conducted over 1000 seconds. A fuel load change was simulated and the chemical conversion was tracked. Over the whole simulation, the attrition and small particles leaving the reactor system was also covered. When comparing the simulation with experiments it could be shown that the gas conversion was sufficiently calculated and the conversion state of the oxygen carrier tracked correctly. The attrition modeling, however, with its parameters measured had to be adapted, so that it fitted the outlet particle flows measured under reaction conditions.

The financial support of DFG (Deutsche Forschungsgemeinschaft) within the priority program SPP 1679: â€œDynamic simulation of interconnected solids processesâ€ is gratefully acknowledged; grant number HA 6935/2-1.

Copyright © American Institute of Chemical Engineers. All rights reserved.

514599 Analysis of FCC Cyclone Fault Diagnosis Technology Based on Particles Information

Abstract:

Cyclones are key equipment in fluid catalytic cracking unit (FCCU), which are used for separating the catalyst particles from oil gas or flue gas. Hence, it is very necessary to carry out condition monitoring and fault diagnosis for the running cyclone. The properties of particles vary at different positions in the particle circulation loop, such as size distribution, morphology and element composition, etc. In consequence, the particles contain much information about operation. Once the cyclone operates abnormally, the information of run-off catalyst particles appear abnormal too, and there exists a corresponding relationship between them. So, taking the catalyst particles as the carrier of fault information, this paper put forward cyclone fault diagnosis technology by extracting catalyst particle information, such as particle packing density, particle size distribution, particle micrographs etc, to build the relationship model between the cyclone fault and catalyst particle information.

Copyright © American Institute of Chemical Engineers. All rights reserved.

514600 The Multi-Hole Throttling Effect and the Erosion Characteristics of the High Pressure Natural Gas

Abstract:

Based on Euler-Lagrange multi-phase flow numerical simulation method, the single hole and multi-hole throttling process of the high pressure natural gas was studied by using the RNG *k-Îµ* turbulent model, the Sutherland Viscosity Law compressible fluid viscosity model, and the advanced erosion model numerical simulation. The calculated mass flow variation law in the throttle choke is comparable with the theoretical calculations, which verified the adopted models and calculating methods. And then the throttling effect and the erosion characteristic of the high pressure natural gas was studied to check its specific law with different throttle holes by using the advanced erosion model. The results showed that, high pressure natural gas can reduce pressure and expansion after flowing through the single hole and multi-hole throttle choke, in which lead a high-speed jet region came up near its outlet. But the maximum jet velocity of multi-hole throttle choke was significantly lower than that of the single hole. At the same time the multi-hole uniform jet may reduce the vortex entrainment and the wall erosion damage, in which can be easier to mix the cold core and achieve stable pressure recovery at the role of the multi-jet mixing effect. Therefore, the multi-hole throttling can effectively restrain the erosion damage and the gas hydrate formation probability.

Copyright © American Institute of Chemical Engineers. All rights reserved.

514601 Numerical Simulation of Flow Field In a Gas Pipe Distributor of the FCCU Regenerator

Abstract:

The flow field in the pipe distributor used in regenerator of FCCU (Fluid Catalytic Cracking Unit), was analyzed based on the numerical simulation. Firstly, the distributing gas performance of the pipe distributor was investigated. The result shows that the pressure increases along the branch pipe of the pipe distributor and the axial flow rate decreases in the branch pipe, which leads to the nozzle flow rate increasing gradually. Furthermore, an obvious bias flow was found in the inlet of branch pipe, which brings some nozzle flow rate less than the designed flow rate and causes an uneven gas distribution in the fluidized bed as well. Meanwhile, the momentum equation with variable mass flow was applied to analyze the flow in the pipe distributor. The flow type in the branch pipe belongs to the "momentum exchange controlling model". The pressure in the branch pipe always increases from the inlet to the end, and the pressure distribution is not consequentially uniform. The uneven gas distribution of the pipe distributor results in fluidized gas non-uniformity in the bed and erosion in the branch pipe and nozzle. Finally, according to the pressure characteristics in fluidized bed system, relationship between the pressure drop of branch pipe and pressure drop of nozzles, as well as of the critical nozzle pressure drop and the designed nozzle pressure drop, were discussed. The gas distributor performance could be optimized by means of the structural improvement of branch pipe.

Copyright © American Institute of Chemical Engineers. All rights reserved.

514602 Full-Loop Simulation of Gas-solids Flow in Circulating Fluidized Bed of FCC System

Abstract:

Circulating fluidized beds are widely used in petrochemical, coal, metallurgy, and other industrial fields. A typical circulating fluidized bed (CFB) includes the riser, gas-solid separator, downer and solids flow control device. To obtain detailed information and get better understand of gas-solids flow in CFB, computational fluid dynamics (CFD) had been applied extensively in many systems such as circulating fluidized bed combustion (CFBC) systems and fluid catalytic cracking (FCC) systems. In most studies, the riser, the core part in CFB system, was isolated as a single simulation object by applying proper boundary conditions. However, there are some limitations of riser-only simulation, without consideration of dynamics in solids circulation rate and effects of other part of CFB system. Therefore, the full-loop simulation is a more promising way to describe complex gas-solids interaction in CFB by treating the whole CFB system as a simulation object. Nowadays, most of current full-loop simulations were focused on CFBC systems, there was little work about FCC system, with different type of particles and recycle mechanism from CFBC system.

In this work, both full-loop and riser-only simulations were carried out by TFM approach to investigate a pilot-scale gas-solid circulating fluidized bed of FCC system which has been set up in our group. The simulation condition was chosen according to experiments with different superficial gas velocity, valve opening and fluidizing air flow rate. Detailed comparison between full-loop and riser-only numerical simulations has been conducted with respect to the hydrodynamics inside the riser, which showed full-loop simulation had advantages on predicting dynamics of solids circulation rate and the gas-solids flow behavior. Distribution of pressure, solids holdup and velocity in system was investigated which was helpful to understand the hydrodynamics in CFB. The effects of operating condition were then studied, which revealed the system stability mechanism.

Copyright © American Institute of Chemical Engineers. All rights reserved.

506758 Immobilization of Sulfur-Oxidizing Bacterium, *Thioalkalivibrio Sp.* D301 on Magnetic Nanoparticles and Biodesulfurization

Abstract:

There is an urgent need to develop desulfurization methods for halo-alkaline wastewater. *Thialkalivibrio sp.* D301, an obligatory chemolithoautotrophic, haloalkaliphilic, sulfur-oxidizing bacterium, was used in the removal of sulfide in wastewater. We developed a process to immobilize D301 cells by using superparamagnetic Fe_3O_4 nanoparticles (NPs) under halo-alkaline conditions. Fe_3O_4 NPs were synthesized by coprecipitation of ferric and ferrous salts in the presence of N_2 gas. TEM image shows the spherical shape of Fe_3O_4 NPs with an average diameter of 10 to 15 nm. The particles had superparamagnetic properties. The saturation magnetization (Î´$_s$) of the Fe_3O_4 NPs was 73.985 emu g$^{\text{âˆ'1}}$ with zero remanence and coercivity. Analysis of zeta potentials and hydraulic diameters showed that the Fe_3O_4 NPs and bacterial cells interact mainly through the nano-size effect. The NPs-coated cells have the same sulfur oxidization activity as that of free cells. Experiment results showed that the immobilized cells still retained >90% of their sulfur oxidization activity after six times of reuse. The cells coated with Fe_3O_4 NPs did not interact through electrostatic attraction. The mechanism of NPs adsorption mainly involved the nano-size effect. Magnetic immobilization is a convenient technique for HA-SOB immobilization under halo-alkaline conditions and is a promising technology for large-scale industrial applications.

Copyright © American Institute of Chemical Engineers. All rights reserved.

507944 Simulation of Bulk Solids and Granular Systems By Using Combined Discrete Element Models

Abstract:

Granular materials can be found in many industrial processes from chemical to energy, agriculture, mining, pharmacy and many other fields. They are almost always opaque, making the acquisition of experimental data below the surface difficult. DEM (Discrete Element Method) simulations are an effective way to study the physics of granular materials because they can simulate the micro-dynamic behavior of granular material on an individual particle scale and the motion of every particle can be traced. Since DEM was first considered, many modified discrete element models have been developed. However, most existing DEM simulation are just for spherical particles or a specific type of non-spherical particles. In this work, a combination of several types of discrete element models including spherical models, super-ellipsoids, multi-sphere models (with no interaction and deformation between sub-spheres), and bonded-spheres models (with interaction and deformation between sub-spheres) were used to simulate the flow of granular materials with many different shapes. Several simulation cases including the flow of particles in tumblers, blenders, and other devices were performed to validate the modeling approaches. The simulation results show that it is practical and efficient to combine different discrete element models together for simulating both spherical and non-spherical particles in bulk solids and granular systems.

Copyright © American Institute of Chemical Engineers. All rights reserved.

515703 Material Optimization of Perovskite Films using High Throughput Synthesis and Multi-Dimensional Analysis

Abstract:

Over the past years, Perovskites received special attention in the field of renewable energy as they are one of the prime candidates for energy storage and conversion materials. Perovskites act as the active light harvesting layer of a solar cell and provide excellent performance.[1] A common synthesis strategy is through the liquid phase under inert atmosphere.[2,3] However, this comes with the challenge of a huge parameter space that in principle would have to be first synthesized, then characterized and finally optimized based on the therefrom deduced knowledge.

To overcome these limitations, we follow a fully automated approach, which allows automated high throughput (HT) synthesis and HT characterization of perovskites with high accuracy. As an efficient route to explore the parameter space, we use design of experiments (DoE) to find optimized settings for process parameters such as synthesis and annealing time and temperature, but also material aspects like the molar ratio of the reactants.

For all experiments, a Chemspeed© technologies (CST) swing XL platform with a liquid handling tool and a heatable shaking rack was used. The robotic platform is enclosed within an Mbraun© glovebox with inert atmosphere (Nitrogen). After precursor preparation, deposition was realized by a liquid handling tool on glass substrates inserted in a customized aluminum 24 well plate, which is then transferred onto a hot plate for film formation.

In a first step, after annealing at different temperatures, a characterization cascade was developed and applied that contains the subsequent elimination of samples that exhibit undesired properties and side products. Fast absorption analysis, photoluminescence (PL) measurements and finally X-ray diffraction (XRD) were performed to identify optimum chemistry and process parameters. Although this is a very reliable procedure, it is still comparatively demanding in terms of the time effort. Therefore a more efficient, purely optical approach by which the experimental parameter space is only explored in terms of rapid emission analysis was developed in a second step. Noteworthy, both strategies led to the same optimum conditions, however, at significantly reduced time effort.

In conclusion, our work shows how by the combination of automation, DoE and efficient characterization, optimum conditions for new materials like perovskites are rapidly identified. This opens the door towards faster material development which is important whenever new perovskite structures, e.g. lead free materials for industrial applications, are strived for.

Literature

1. Green, M.A.; Ho-Bailie, A.; Snaith, H.J., "The emergence of perovskite solar cells", Nature Photonics 2014, 8, (7), 506-514.
2. Baikie, T.; Fang, Y.; Kadro, J.M.; Schreyer, M.; Wei, F.; Mhaisalkar, S.G.; Graetzel, M.; White, T.J., "Synthesis and crystal chemistry of the hybrid perovskite (CH 3NH3)PbI3 for solid-state sensitised solar cell applications", J. Mater. Chem. A 2013,1 , 5628-5641.
3. Yan, K.; Long, M.; Zhang, T.; Wei, Z.; Chen, H.; Yang, S.; Xu, J., "Hybrid Halide Perovskite Solar Cell Precursors: Colloidal Chemistry and Coordination Engineering behind Device Processing for High Efficiency", J. Am. Chem. Soc. 2015, 137, (13), 4460–4468.

Copyright © American Institute of Chemical Engineers. All rights reserved.

509509 Solid Liquid Separation Via Particle Flow Instability

Abstract:

A new separation strategy using novel particle-flow-instability physics is successfully developed for a *difficult-to-separate suspension* in which there is some combination of a small density difference between solid and liquid, high viscosity, and small-sized particles. The method we propose here requires no dilution for medium, high or extremely high viscosity slurries and can produce effective solidâ€“liquid separation with low capital and operational inputs. The results have relevance for many biotechnological and/or chemical processes.

Copyright © American Institute of Chemical Engineers. All rights reserved.

509543 Milling and Grindability Assessment of Pharmaceutical Materials

Abstract:

Milling is one of the most important unit operations in the pharmaceutical industries. Most of the Active Pharmaceutical Ingredients (APIs) and excipients are produced by crystallisation, and grown to a sufficiently large particle size suitable for isolation and drying. Therefore most of the materials have to be milled to desirable particle sizes. Milling is influenced by a combined effect of milling dynamics (input power, mill geometry, etc) and material properties. Considerable research has been carried out on analysing the particle behaviour in various mills. Nevertheless, the prediction of the milling rate and milled product properties (e.g. product size distribution) is still very challenging.

In this work, the milling behaviour of several pharmaceutical materials in the PicoPlex mill (pin mill unit of the Picoline, Hosokawa Alpine) has been studied. A combination of experiments and Distinct Element Method (DEM) simulations is used to explore the dynamic behaviours of the materials in the mill. The results are then used to validate a novel approach of grindability assessment of pharmaceutical materials using the Scirocco disperser of the Malvern Mastersizer 2000 (Bonakdar et al. 2016). Based on this method the ease of grindability of the materials in the mill can be predicted.

Bonakdar, T., Ali, M., Dogbe, S., Ghadiri, M., and Tinke, A. (2016). A method for grindability testing using the scirocco disperser. *International Journal of Pharmaceutics,* 501: 65-74

Copyright © American Institute of Chemical Engineers. All rights reserved.

507857 Bi-Directional Thermal Control of Twin Screw Granulation Process Via a Specialised Annular Heat Pipe

Abstract:

The continuous process of twin screw granulation forms a hotbed in pharmaceutical research as it is expected to succeed the current batch production of granules. In essence, the scale up of the material output is a function of production time, instead of equipment size (for batch granulators). This way, a â€œjust-in-time productionâ€ of the desired amount of material can be performed according to specific needs. In this work, a twin screw granulator was utilised in the granulation of both the placebo and active pharmaceutical formulations. Subsequently, the same equipment was used in granulating ceramic raw material powders for tile pressing. During processing, the spatial constraint in the barrel and screw compartments caused friction, resulting in excess heat generation. For the thermally-sensitive pharmaceutical formulations, this must be controlled to avoid denaturation of the granule products. A common solution is by using flowing cool water, but a good control has been proven to be difficult to achieve as different granulation zones generate different heat levels (kneading blocks > conveyer blocks). Heat pipe is a passive and highly effective thermal conductor and stabiliser. A carefully-designed annular heat pipe, filled with a small amount of methanol was successfully employed to stabilise the temperature in the barrel space, cooling the granulating material before it reached its critical temperature. For the ceramic powders, the methanol filling of the heat pipe was replaced by water and it was then heated up to provide an in-situ drying operation as the powders granulating. The in-situ drying in this instance, reduces energy consumption for the downstream drying processing as the product granules contain less moisture to be removed. The annular heat pipe, with its bi-directional operation of heating and cooling is a technically viable temperature controller for twin screw granulators. Compared to the established active cooling methods, the heat pipe provides a passive alternative, with only a discreet amount of fluid required to last its operational lifetime, thereby providing an economic thermal control solution.

Copyright © American Institute of Chemical Engineers. All rights reserved.

Mixing Grains with Different Elongation in a Rotating Drums

Claudia Piacenza[1,2], *Marco Marconati*[1], *Colin Hare*[1], *Andrea C. Santomaso*[2], *Marco Ramaioli*[1,*]

[1]*Department of Chemical and Process Engineering, Faculty of Engineering and Physical Sciences, University of Surrey, Guildford GU2 7XH, UK*
[2] *APT Lab, Department of Industrial Engineering, University of Padova, Padova, Italy*

[*] *Email: m.ramaioli@surrey.ac.uk*

Abstract

Mixing grains with different properties is a remarkably challenging process, relevant to many industrial applications. Rotating drums have been used extensively as model systems to study granular media flow and mixing and segregation. Numerous studies considered the mixing of grains with different sizes, but only few studies considered shape and elongation, which have been already identified as important characteristics affecting packing and inducing segregation. In this contribution, the mixing of binary mixtures of grains having the same volume, but different elongations was studied experimentally. The mixing dynamics of a layered granular medium was characterized in a rotating drum, highlighting the impact of the grain shape, of the drum angular speed and of the drum filling ratio. A mixed or segregated state is reached very rapidly, but experiments are continued to verify that the state reached is a steady state. The experiments demonstrated that below a critical difference of elongation, the grains can be mixed effectively. Conversely, when grain elongations are very different, a central core is formed, rich in the more elongated grains. In this case only limited mixing can be achieved. These results can guide the formulation of mixtures of grains with different elongations.

1. Introduction

Mixing is a process operation aimed at obtaining a homogeneous system starting from two or more components. On the contrary, when there is a lack of homogeneity within the mixture segregation occurs. Segregation phenomena represent a relevant problem in the case of mixing solid particles having different physical properties such as particle size, density, shape or surface roughness. Differently from liquids and gases, solid particles are not thermal activated system [1] and so in order to obtain a certain degree of mixing they need to be mechanically activated. Typically three mixing mechanisms are largely recognized:

- dispersion, analogous to diffusion in fluids, it involves single particles that randomly move due to collision between each other;
- convection, it involves a portion of the bulk material, with a given direction and velocity;
- shear mixing: it is peculiar to granular solids and it is a combination of the two previous mechanisms.

Powder mixers are relatively simple machines of low capital cost. They are classified into two types: tumbling mixers and agitated mixers. Tumbling mixers are the simplest. A totally enclosed vessel is rotated about an axis causing the particles within the mixer to tumble over each other on the mixture surface. Tumbling vessels provide low-shear environments. Convective mixers instead are characterized by the action of impellers or paddles to move the powders around and generate a well-mixed product. Convective mixers provide more shear into the mixture and can be used with cohesive materials.

Rotating drums belong to the tumbling mixers. They are extensively used in the chemical and process industries. With respect to mixing processes, rotating drums are often used to study and understand mixing and segregation phenomena; as a result, many experimental and numerical studies were developed on granular behavior in rotating drums [1-3]. Different granular flow regimes can occur in the drum, each one with its specific flow behavior. In particular, different flow regimes can be assessed by using the dimensionless Froude number, $Fr = \omega^2 R/g$ [4], where ω [s^{-1}] is the drum rotation speed, R [m] the drum radius and g [m/s^2] the gravitational acceleration. With $10^{-4}<Fr<10^{-2}$ and filling level $\varphi>0.1$ rolling regime takes place. This motion is characterized by a uniform, steady flow of a particle layer on the surface (cascading or active layer) while the larger part of the bed (plug flow region) is moved upwards by solid rotation with the rotational speed of the wall. A uniform and good mixing is achievable with this type of motion, but when particle size distribution is very broad segregation can appear. The rolling bed is preferred, since is the simplest regime providing favorable conditions for heat transfer and high quality of the product even when mass flow rates are large.

When particles to be mixed have different properties such as different size, different density or different shape segregation typically occurs and particles tend to separate. [5, 6]. Difference in size is by far the most important cause of particle segregation. Particle size distribution influences also strongly bulk properties such as packing fraction and bulk modulus [7].
Different size segregation mechanisms can be identified:

- trajectory mechanism: according to the Stokes' law the velocity of fine particles is lower than that of big particles. This can cause segregation when small particles move through air;
- percolation mechanism: if a bulk material is disturbed, a rearrangement in the particle packing occurs and for fine particles it is easier to fall down in the inter-particle voids;
- inertial mechanism: in pouring and filling operations bigger particles tend to flow selectively to the bottom of the pile since they have a higher momentum;
- elutriation mechanism: in presence of air the very fine particles can remain in suspension while larger ones tends to settled. This can occur in charging storage vessels or hoppers.

A homogeneous binary granular mixture, with differences in size and density, in a two-dimensional rotating drum typically gives radial segregation. Considering a rotating drum operating in the rolling regime filled with particles of different sizes, bigger particles mainly tend to the edge of the system, while smaller particles accumulate to the centre of the drum. The phenomenon is fast and the rotational speed of the drum seems not to affect the phenomenon; the size ratio of the particles is, instead, the parameter that mainly affects size segregation phenomena. Most of the experimental investigations about segregation in the rotating drum involved spherical or nearly-spherical particles; the experimental studies of the flow of non-spherical particles are, in fact, rather limited [6,8]. The shape of the particles, however, strongly influences the motion in the system. In particular, the static and dynamic angles of repose of non-spherical particles are found to be very different from the ones measured for spherical particles suggesting that differences in particle mobility can affect the segregation process [9].

2. Materials and methods

2.1 Particles

Both spherical and non-spherical particles were used for the experiments. Their size (diameter d for spheres, major and minor axis for ellipsoids a and b respectively) , aspect ratio AR, projected area A_p, volume V, and true density ρ are reported in Table 1.

Table 1. Physical properties of the particles

Type	d [mm]	a [mm]	b [mm]	AR [-]	A_p [mm^2]	V [mm^3]	ρ [g/cm^3]
Spheres S	8	-	-	1	50.3	268	1.09
Ellipsoids E1	-	10.48	6.98	1.5	57.5	267	0.96
Ellipsoids E2	-	18.44	5.26	3.5	76.2	267	0.98

Spherical particles with size 8 mm were used for the experiments. The ellipsoids were 3D printed with *AR* equal to 1.5 and 3.5. They were printed with the Ultimaker 3D printer and made of polylactic acid (PLA). The software used to realize the 3D design of the particles used in this work is FreeCAD, which is an open-source parametric 3D modeller allowing to obtain STL files.

The true density of all the particles used for the experiments was measured by gas pycnometry (Multivolume Pycnometer 1305, Micromeritics). The measured density of the spheres was very close to the typical density of the nylon 66 (1.14 g/cm^3). The PLA ellipsoids had a density that was lower than the typical density of the PLA (1.24 g/cm^3). This is because the particles were printed with a set infill density of the 20%, so they where partially empty inside.

2.2 Tumbling mixer

The quasi-2D drum was made of two transparent polycarbonate disks with a diameter of 30 cm and a thickness of 6 mm; between them a wooden spacer was placed, with the same external diameter, an internal diameter of 25 cm and a thickness of about 9.1 mm. The spacer allowed to create an internal volume of finite thickness separating the two polycarbonate disks.

a) b)

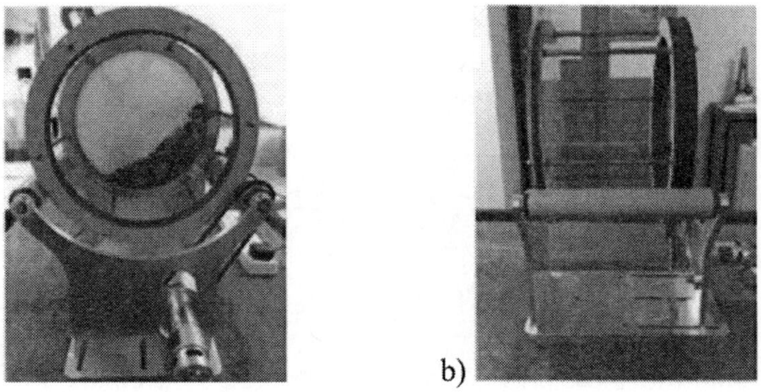

Figure 1. a) Front view and b) side view of the rotating drum (the drum is on the left side of the rigid structure).

The internal lateral wall of the volume was covered with sandpaper in order to provide friction and avoid undesired slip of the particles. To keep the drum in vertical position and allow its rotation, three additional wooden spacers were used, at a distance of about 16.3 cm. This rigid structure could rotate at variable rotational speed thanks to a motor connected to a voltage supplier. Two rollers mounted on a rigid support made the rotation of the drum possible; a pulley transmitted the motion to the driving roller, while the second roller was passive. The rotational speed could be changed by varying the motor voltage. The structure could be removed from the rollers and by unscrewing the bolts, the drum could be easily opened so that to load/unload the particles.

2.3 Image acquisition

Frontal pictures of the drum were taken by a color camera and analyzed by the image processing software ImageJ [10]. To avoid the reflection of the camera and other objects on the front transparent disk, a squared LED panel was used which allowed to have a dominant diffused light illuminating the drum. The LED had a power of 8 W and a color temperature of 4000 K; it was placed behind the drum and opposite to the camera. A circular sheet of paper, which was attached to the drum between it and the LED, diffused the light properly. The camera was a Basler ace camera – model acA2000-165uc, with a maximum resolution of 2040 × 1086 pixels. The focus and the diaphragm of the camera could be adjusted manually.

2.4 Experimental plan

The attention was focused on mixtures of particles with different shapes, since segregation phenomena due to size had been widely investigated in the past. Three mixtures were considered, with two possible initial top-bottom segregated configurations. Filling level and rotational speed where changed also as detailed in Table 2.

Table 2. Experimental tests performed with differently shaped particles.

AR 1 [-]	*AR* 2 [-]	ω [rpm]	φ [-]	Init. Config.
1	1.5	4.22	25%	E1/S
1	1.5	4.22	25%	S/E1
1	1.5	4.22	47%	E1/S
1	1.5	4.22	47%	S/E1
1	1.5	11.51	25%	E1/S
1	1.5	11.51	47%	E1/S
1	3.5	4.22	25%	E2/S
1	3.5	4.22	25%	S/E2
1	3.5	4.22	47%	E2/S
1	3.5	11.51	25%	E2/S
1	3.5	11.51	25%	S/E2
1	3.5	11.51	47%	E2/S
1.5	3.5	4.22	25%	E2/E1
1.5	3.5	4.22	25%	E1/E2
1.5	3.5	4.22	47%	E2/E1
1.5	3.5	4.22	47%	E1/E2

*AR*1 and *AR*2 indicate the aspect ratios of the particles in the binary mixture. It is reminded that the spheres (*AR*=1) and the two types of ellipsoids (*AR*=1.5 and *AR*=3.5) had the same volume. The E1/S initial configuration indicates a segregated configuration with short ellipsoids above spheres (this was assumed as basis configuration); the S/E1 configuration is inverted with spheres above short ellipsoids. A similar convention is applied to the experiments considering mixtures of ellipsoids. All the experiments were performed with binary mixtures with a composition of 50% and 50% on a number (and hence also volume) basis. The filling level φ of the 25% was obtained with a total number of particles equal to 200; with 320 total particles the filling level of the drum was equal to 47%.

2.5 Image analysis procedure

Image analysis techniques were used to identify the particles on the basis of their different shape. The procedure was automated using customized ImageJ macros. Contrast and brightness of the images were first adjusted. Color images were then set to gray scale and segmented by applying a threshold. The operation was carried out twice to separate light and dark particles with different thresholds. Watershed segmentation was then used to separate touching objects. This last operation was more difficult for the particles at the border of the drum; in some case, in fact, the ellipsoids were erroneously identified as spheres. Also, the presence of the border of the drum complicated the watershed operation.

The centre of mass of each particle ($x_{m,i}$ and $y_{m,i}$) was estimated and the radial distance (d_i) from the centre of the drum (x_c, y_c) calculated as:

$$d_i=[(x_c-x_{m,i})^2+(y_c-y_{m,i})^2]^{0.5}$$

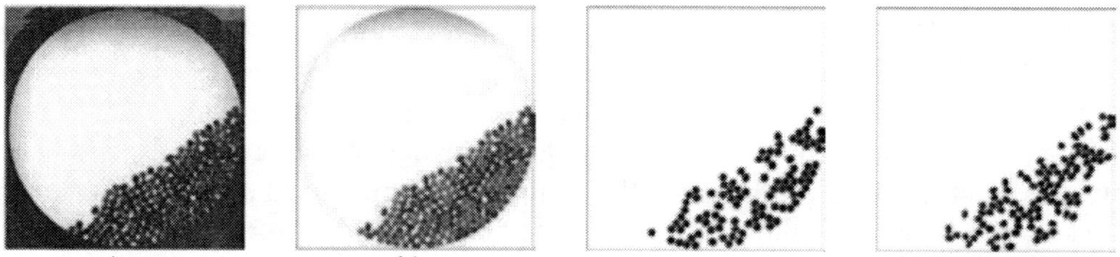

a) b) c) d)

Figure 2. Example of image processing steps. a) Original colour image; b) grayscale conversion and circle cut from the original picture; segmentation applied to c) light particles and d) dark particles.

The mean value of distance (d_m) for the spheres and for the ellipsoids was then calculated as:

$$d_m = \Sigma\, d_i/N$$

Only for very elongated ellipsoid (*AR*=3.5) the watershed operation failed resulting in over-segmentation errors. Single particles could not be identified unequivocally in this case; only one center of mass for all the elongated ellipsoid was estimated and the average distance (d_m) of the particles from the drum center calculated.

3. Results and discussion

Figure 3 shows two typical experiments, carried out at the lowest rotational speed and filling ratio, with ellipsoids (*AR*=1.5 at left and *AR*=3.5 at right) on top and spheres on bottom in the initial and final configuration. It is evident even by visual inspection that the final state presented some level of radial segregation and that a different final segregation intensity was reached in the two systems.

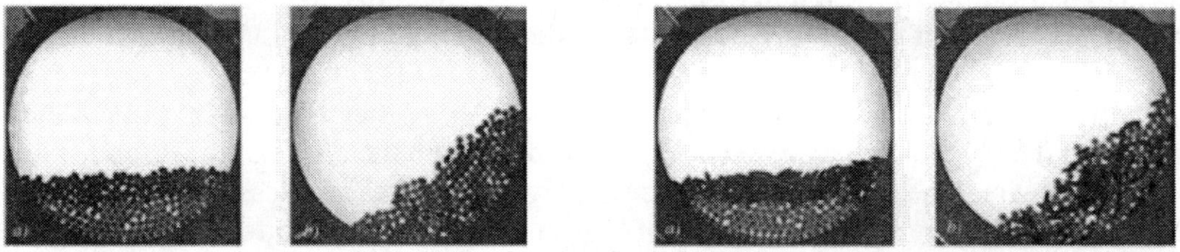

Figure 3. Drum at a) t=0 and b) t=30 min. On the left the initial configuration is E1/S (using ellipsoids with *AR*=1.5) and on the right the initial configuration is E2/S (*AR*=3.5). In both experiments ω=4.22 rpm and φ= 25%.

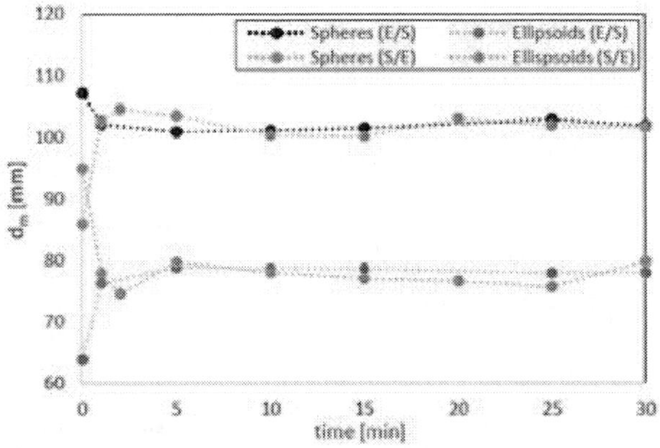

Figure 4. Evolution of the mean radial distance vs. time for the mixture of spheres and ellipsoids with *AR*=3.5. The initial configuration E2/S is compared to the initial configuration S/E2.

A more quantitative analysis was obtained by plotting the trend of the mean radial distance (d_m) of the two species at different time instants. As a typical example of the system evolution, the case of E2/S

is shown in Figure 4. The d_m reached a steady value after a short time. Dedicated experiments on a shorter time scale (2 minutes, not shown here) revealed that steady state value for d_m was reached after 30s for mixtures with ellipsoids E1 (AR=1.5) and after 60s for mixtures with E2 (AR=3.5).
At steady state, the radial distance was in both cases smaller for the ellipsoids, meaning that they mainly remained concentrated on top or at the center of the granular bed while the spheres mainly concentrated at the periphery of the drum (close to the drum wall). This tendency was much stronger for ellipsoids E2. So a slower segregation dynamics compensated by a larger segregation intensity was observed with the more elongated particles.
As Figure 4 shows, inverting the initial configuration from E/S to S/E did not affect the dynamics and the final result in all the experiments performed with these particles. Initial configuration resulted to have no effect on the segregation propensity of the mixtures analyzed.

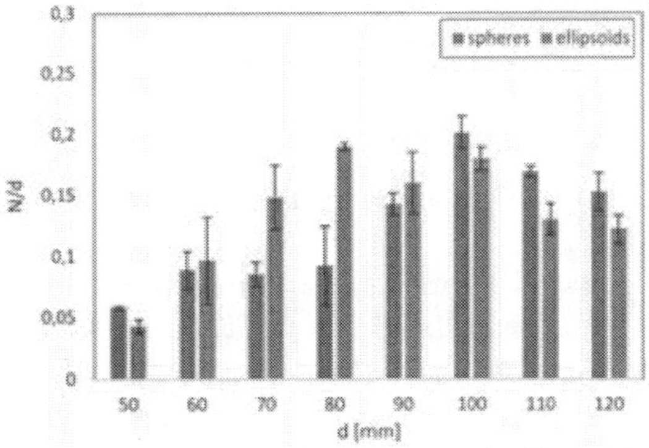

Figure 5. Distribution of the radial distance of spheres and ellipsoids with AR=1.5, obtained by averaging the results of three repetitions. Other conditions are t=30 min, ω=4.22 rpm, φ= 25%.

A more in-depth analysis can be carried out considering the distribution of the radial distance of the spheres and ellipsoids separately. The radial distance from the centre of the drum (d) was divided in classes of 10 mm and the number of each type of particles (N) in each class was computed. To obtain the distribution, N was normalized on the radial distance as N/d and plotted versus d. In the case of completely homogeneous mixtures and assuming that the medium takes the shape of a circular sector, N/d should be constant and equal for the two types of particles in each class. The peak of the distribution of the ellipsoids was found at lower distance than the peak of the distribution of the spheres. Moreover, at lower distance, the number of ellipsoids was higher than the number of spheres, especially for distances between 70 and 90 mm; at higher distance (closer to the drum wall), instead, the number of spheres was higher than the number of ellipsoids, thus confirming the previous analysis based on d_m.

Also, binary mixtures made of the two types of ellipsoids E1 and E2 (AR=1.5 and AR=3.5 respectively) were taken into account at the same filling levels of the previous experimental tests and at the lower rotational speed. Stable segregation was again observed with intermediate results with respect to the two previous cases (S+E1 and S+E2). So as a general comment on these experiments, stable segregation was always observed when mixing particles having same volume but different AR. The intensity of segregation increased with increasing the difference in AR between the particles. In particular particles with larger AR tended to accumulate at the center of the granular bed, while particle with lower AR accumulated at the periphery. Segregation occurred in the surface active layer where the more elongated particles, because of their reduced propensity to flow (particle rolling was prevented by their elongated shape), were systematically overtaken by the less elongated (and faster) particles. In this way a spontaneous separation of particles occurred in the active layer, with those having a lower AR arriving first at the drum wall, while those having an higher AR remaining trapped in the core.

3.1 Effect of filling level

The comparison of the behavior of the mixtures at the two filling levels can be carried out by normalizing the mean radial distance d_m of the two types of particle on the mean distance of all the particles d_c, calculated as $(d_{m,\ spheres}+d_{m,\ ellips.})/2$. The trends of d_m/d_c are shown in Figure 6

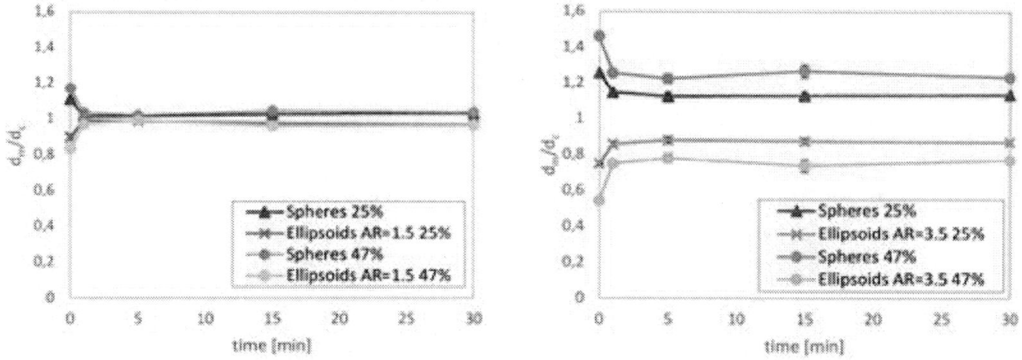

Figure 6. Trend of d_m/d_c vs. time for the mixture of spheres and ellipsoids with AR=1.5 (left) and AR=3.5 (right) at two different filling levels.

In the case of mixture of spheres and ellipsoids with AR=1.5, the curves at different filling levels overlap, showing that the behaviour of the mixture was not affected by the filling level of the drum. In the case of the other two mixtures (S+E2 and E1+E2, not shown), the curves did not overlap , especially for the case S+E2 shown in Figure 6 (right). An explanation can be found in terms of bed geometry. When increasing the filling level (below 50%) the length of the surface active layer increased as well. This gave more chances for particles to separate, in particular when the difference in AR was large enough (S+E2 and E1+E2). However for the system S+E1 where the difference in AR was the smallest, the difference in active layer length in the two cases was not sufficient to create observable differences in the segregation pattern.

3.2 Effect of rotational speed

In addition to experiments carried out at 4.2 rpm, tests were repeated at 11.5 rpm with spheres and ellipsoids (25% filling level). Figure 7 shows that rotational speed did not affect the segregation patterns, at least in the range of the analyzed speeds.

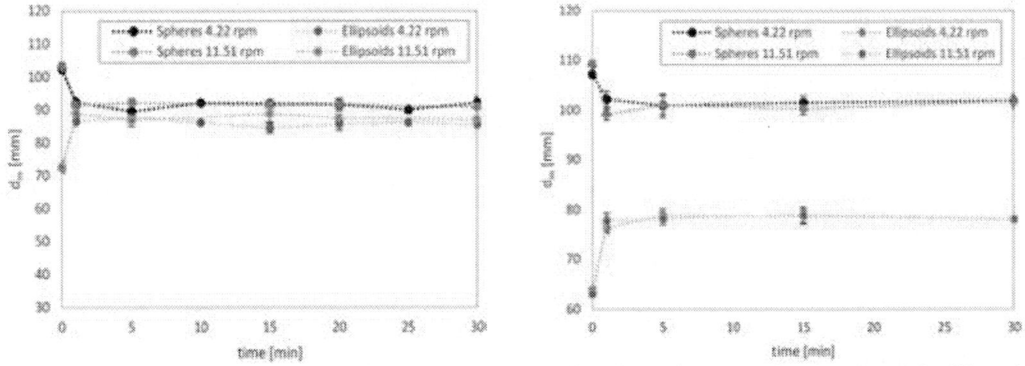

Figure 7. Trend of the mean radial distance vs. time for the mixture of spheres and ellipsoids with AR=1.5 (left) and AR=3.5 (right) at two different rotational speeds.

Conclusions

From the experimental tests described in the previous paragraphs, it was found that binary mixtures of particles with different shape but with the same volume do not mix uniformly and segregation was always observed. In particular, the particles with lower aspect ratio tended to stay closer to the drum walls, while the more elongated particles were found preferentially at the centre of the granular bed. The most significant case was represented by the mixture of spheres and ellipsoids with *AR*=3.5; for this mixture, in fact, the segregation was more evident and the non-spherical particles clearly accumulated in the centre of the granular bed. The same final state was obtained regardless of whether the longer particles were initially located to the top or the the bottom of the mixture. A proportionality between *AR* difference and segregation intensity was observed. Some variables such as rotational speed and initial bed configuration were found to not affect the segregation dynamics and intensity. Only filling level (in addition to *AR*) was found to affect segregation intensity. A mechanistic explanation was given in terms of differences in particle flowability occurring in the surface active layer (confirmed also by an analysis of the angles of repose in the drum). In this specific case therefore percolation and inertial effects (typical of size segregation) were not the main driving mechanisms for the observed segregation (the particle indeed had the same volume and similar mass so that sterical and inertial phenomena were minimized). Segregation mainly occurred because of differences in particle velocity in the active layer due to a reduced mobility of particles with larger *AR*.

References

1. GDR MiDi (2004), "On dense granular flows", The European Physical Journal E, 14, 4, 341–365.
2. Jarray A., Magnanimo V., Ramaioli M., Luding S. (2017) Scaling of wet granular flows in a rotating drum, EPJ Web of Conferences, 140,03078.
3. Volpato S., Canu, P., Santomaso A.C. (2017), "Simulation of free surface granular flows in tumblers", Advanced Powder Technology 28(3), 1028-1037.
4. Mellmann J. (2001), "The transverse motion of solids in rotating cylinders – forms of motion and transition behaviour", Powder Technology, 118, pp 251-270.
5. Windows-Yule C.R.K., Scheper B.J., van der Horn A.J., Hainsworth N., Saunders J., Parker D.J., Thornton A.R. (2016), "Understanding and exploiting competing segregation mechanisms in horizontally rotated granular media", New Journal of Physics, 18, 2, 023013, 2016.
6. Santomaso, A.C., Olivi, M., Canu, P. (2004), "Mechanisms of mixing of granular materials in drum mixers under rolling regime", Chemical Engineering Science, 59 (16), 3269-3280.
7. Kumar N., Magnanimo V., Ramaioli M., Luding S. (2015), "Tuning the bulk properties of bidisperse granular mixtures by small amount of fines", Powder Technology, 293, 94-112.
8. Ramaioli M., Pournin L., Liebling Th. M., (2005), "Numerical and experimental investigation of alignment and segregation of vibrated granular media composed of rods and spheres", Proceedings of the 5th International Conference on Micromechanics of Granular Media, 2, 1359-1363.
9. Santomaso, A.C., Canu, P. (2007), "Single particles properties vs. bulk flowability", Proceedings of 5th International Conference for Conveying and Handling of Particulate Solids.
10. Schneider, C. A., Rasband, W. S., Eliceiri, K. W. (2012), "NIH Image to ImageJ: 25 years of image analysis", Nature methods 9(7): 671-675.

507441 Generation of Particles with a Special Morphology By Desublimation of Copper Phthalocyanine

Abstract:

The morphology of submicron particles becomes more and more important concerning new functional materials. For example the special electric properties of particles can be used in electronic devices [1]. Moreover the electric, optical or catalytic properties of particles can be optimised by adjusting the morphology of the particles. This may imply a risk for the health but makes it possible to realize specific product characteristics. The morphology of the particles should therefore be considered e.g. for filter tests and for the developing and testing of sensor systems.

To explore the characterization of the particle morphology the DFG-AiF cluster project "multi-parametric characterization of particle-based functional materials using innovative online measurement systems" (MPaC) includes the development of online sensor systems for the size and structure of the particles as well as the development of generation methods for particles with a special morphology. The production of spherical particles can be achieved with different commercial aerosol generators. Several generation techniques for special particles are described in literature [2] [3] [4]. To test the sensor systems with particles of different shapes a new type of aerosol generator was developed. The aim for the new generator is to produce the particles in the gaseous phase and provide the particles directly for sensor or filter testing.

The developed generator uses the desublimation of copper phthalocyanine (organic pigment) to structure the surface of titanium dioxide. The bulk material copper phthalocyanine is dispersed into a nitrogen gas stream and transported to a tube furnace. Here the bulk material sublimates and the gas stream carries the material to a laval nozzle where the adiabatic expansion takes place. This leads to a rapid decrease of the temperature behind the laval nozzle and an oversaturation in the gaseous phase. In addition to the decrease of the temperature core particles of titanium dioxide were initiated behind the laval nozzle. The copper phthalocyanine desublimates on the core particles and causes a structured surface. The morphology of the generated particles is characterized by SEM. The particles show a punctual coating of the titanium dioxide particles with copper phthalocyanine. The size of the titanium dioxide particles is about 500 nm and the punctual coating of copper phthalocyanine measures a few nanometres.

The generation method and the design of the generator will be presented and the morphology of the generated particles will be described.

[1] K. Deppert, M. H. Magnusson, S. B. Carlsson, B. Gustafson, T. Junno, . L. Montelius, B. J. Ohlsson, L. Samuelson und L. E. Wernersson, „Future electronic devices by aerosol techniques," *Journal of Aerosol Science,* Nr. 31, 2000.

[2] J. Raula, A. Kuivanen, A. Lähde, H. Jiang, M. Antopolsky, J. Kansikas und E. . I. Kauppinen, „Synthesis of image-leucine nanoparticles via physical vapor deposition at varying saturation conditions," Journal of Aerosol Science, Nr. 38, 2007.

[3] M. Heurlin, M. H. Martin, D. Lindgren, M. Ek, R. L. Wallenberg, K. Deppert und L. Samuelson, „Continuous gas-phase synthesis of nanowires with tunable properties," Nature, Nr. 492, 2012.

[4] K. Nielsch, K. Deppert, M. H. Magnusson, G. Dumpich, E. Kruis und H. Fissan, „Formation of ultrafine particles from powders of compounds with incongruent sublimation behavior," Journal of Aerosol Science, Nr. 28, 1997.

Copyright © American Institute of Chemical Engineers. All rights reserved.

516184 Rubbery Milling of Seed Endosperms for Improved Sustainability by Natural Functionality Preservation

Abstract:

Increasing demand for functionalised goods addressing individual needs requires a profound understanding of process-induced alterations. Development of models and process-structure-property relationships are essential to control the resulting quality and properties of processed food. A wide range of food systems contain solid or semi-solid particulates as major structure building blocks. Generation and tailoring of such particulates strongly depends on their size distribution, shape, morphology, phase behavior (amorphous/crystalline), interfacial properties and related inter-particle force interactions.

A novel processing technology was developed widening the frame of possible alterations, aiming towards tailoring such particle properties generating novel food structures or preserving techno-functional characteristics of the raw material. The technology employs targeted adjustments of the initial moisture content in combination with milling temperature. The resulting transition from the glassy to the rubbery state generates different material properties and processing behaviour. The technology is highly versatile, owing to the universality of moisture induced state transitions.

Seeds from locus bean gum or guar gum were wetted with excess water in order to adjust a rubbery state of the seed endosperm material. In this state milling within a novel type of impact mill with superimposed drying and sifting steps causes more gentle material disintegration compared to conventional dry milling. This preserves the high molecular weight fraction of e.g. galactomannans (micro-scale structure) from degrada-tion and allows at the same time the adjustment of a micro-porous morphology (macro-scale structure) improving wetting, water uptake and dispersing characteristics of the resulting thickener powders. Such structure preservation /modification increased the thickening characteristics extraordinarily as will be demonstrated.

From this prominent rubbery milling example we have entered in related systematic work addressing a wider range of seed endosperms with various â€œnature-synthesizedâ€ func-tionnalities to be further addressed.

/1/ Pollard, M., Eder, B., Fischer, P., and Windhab, E. J. (2010) "Characterization of galacto-mannans isolated from legume endosperms of Caesalpinioideae and Faboideae subfamilies by multidetection aqueous SEC." *Carbohydrate Polymers*, 79(1), 70-84.

/2/ Illmann, S., Stoob, M., Sieber, C., Konz, A.K. and Windhab, E.J. (2012) "New Thickening Quality of Galactomannan Polysaccharides by Tailored Impact Milling" *XVIth Int. Congress on Rheology*, Lisbon, Portugal, August 5-10

Copyright © American Institute of Chemical Engineers. All rights reserved.

503422 Particle Technology Education at Purdue University

Abstract:

Degree programs and coursework at Purdue University focused on particle technology will be presented. Additional information may be found at: https://engineering.purdue.edu/CP3

Copyright © American Institute of Chemical Engineers. All rights reserved.

506453 Teaching Particle Technology in Portugal â€“ University of Coimbra

Abstract:

Education in Particle Technology started in Portugal in the 80â€™s (last years of the decade). The Chemical Engineering degree of the University of Coimbra was the first engineering degree, in the country, to include this topic in the curriculum. Particle Technology was recognized by us, at that time, essential for the education of chemical engineers, in parallel to education in the traditional areas of Chemical Engineering which were still focused on monophasic systems.

Since then, Particle Technology has always been present in the Chemical Engineering curriculum of the University of Coimbra, in different configurations. During the first years particle characterization was combined with design of operations involving particulate systems (grinding and pneumatic and hydraulic transport). From 2005 to 2010, with the curriculum reformulation in relation with the Bologna Process, the two topics started to be addressed in different courses: one course on Materials Characterization and the process topic in a course entitled Structuring and Stabilization Processes, dealing with colloids and emulsions, grinding, agglomeration and spray drying and, for the first time, an integrated modelling strategy for particulate processes based on the population balance formulation was introduced in this course. The two courses were compulsory, at this time.

From the school year 2011/2012 till present, the configuration of this area in the Chemical Engineering degree of the University of Coimbra changed again, even if not deeply (see https://apps.uc.pt/courses/EN/course/361). The course on materials characterization, which includes general notions on particle characterization is kept as an autonomous course included in the 1st cycle of the Chemical Engineering degree of the University of Coimbra. A second course on design of processes involving particulate systems now exists in the 4th year of the MSc degree (1st year of the 2nd cycle), entitled Disperse Systems Technologies, which is an optional course, and which deals with all types of disperse systems, i.e gas, liquid or solids as disperse phase. This course starts with a review of the mechanisms/interactions occurring at the interface with the disperse phase, goes on, again, with a chapter on population balances modelling, and proceeds with strategies for disperse systems stabilization, linking this subject with formulation concepts, ending with the design of some processes involving particulate systems, such as colloids and emulsions, crystallization, grinding, agglomeration and spray drying. This course is also offered to other engineering degrees, such as Materials Engineering and Biomedical Engineering.

Also at the PhD level a course on Disperse Systems Technologies can also be offered during the formal lectures period of the PhD program on Chemical Engineering (or others), depending on the students background and interests.

Since the introduction of the Particle Technology topic in the curriculum of the Chemical Engineering degree of the University of Coimbra, the feedback from employers has been very positive, and the competences in this field acquired by our students have been quite valued by the job market. Till today, the Chemical Engineering degree of the University of Coimbra is still the one, in Portugal, with a stronger emphasis on this topic in the curriculum. In our curriculum, the topic of Particle Technology interacts directly with another course on Product Design, also in the 4th year of the MSc degree and, of course, also with the Project Design course (5th year of the MSc degree), besides other more traditional courses in a Chemical Engineering degree such as in Reaction Engineering and with courses related to environmental technologies.

At present, we have started developing multimedia tools for the teaching/learning of this topic, especially in the field of particle characterization, to be made available on-line, as is the case of the interactive video on nanoparticles characterization (http://labvirtual.eq.uc.pt/siteJoomla/index.php?option=com_content&task=view&id=277&Itemid=470, in Portuguese). Also, the European Federation of Chemical Engineering (EFCE) is considering investing in this type of tools for the field of particle technology, as the result of a collaboration between the Working Groups of Characterization of Particulate Systems and Education.

Acknowledgement

The authors would like to acknowledge the financial support of Pest/C/EQB/UI0102/2013 from FCT/MCTES (PIDDAC), co-financed by the European Regional Development Fund (ERDF) through the program COMPETE (POFC).

Copyright © American Institute of Chemical Engineers. All rights reserved.

506629 Highlights of Particle Technology Teaching in Singapore

Abstract:

Our approach to education in particle technology is to incorporate elements of our research in this field, relevant industrial case studies and apply a Technology Enhanced Learning (TEL) approach for implementing pedagogical interventions to enhance the learning experiences of students.

For supercritical CO_2 processing, case studies in extraction and drying are used to teach different chapters in the curriculum such as:

- Particle characterization
  - The bulk density, size, shape of real industry samples are given to students for analysis using software such as ImageJ
- Flow through packed bed
  - The pressure drop and scale-up of packed bed design for supercritical CO_2 flow through packed bed are used as a worked example and assignment problem to help students appreciate the role of density and viscosity in the pressure drop in a packed bed
- Drying
  - Supercritical CO_2 as an alternative drying media as compared to hot air or superheated steam. Case studies and examples provided

For gasification and chemical looping combustion applications, case studies in the treatment of wastes and carbon capture are similarly used as part of the syllabus:

- Particle characterization
  - The effect of polydispersity on various fluidization characteristics (entrainment, transport dis-engaging height, clusters, segregation, etc)
  - The effect of agglomeration
- Reliability of correlations and simulations
  - Correlations in literature give predictions orders-of-magnitude apart
- Scale-up
  - Rule-of-thumbs in scaling up the various regimes

A design software has been developed within the paradigm of Technology Enhanced Learning (TEL) to facilitate learning via a repeated practice approach. The software generates detailed solution steps to typical engineering design problems encountered within this module. Students can compare the complete solutions generated with their own hand calculations, and thereby apply a repeated practice approach towards their learning of engineering design calculations. The software has been successful in enhancing the performance of students in the final examination.

Copyright © American Institute of Chemical Engineers. All rights reserved.

507116 Particle Technology Course at the University of Salerno

Abstract:

The Course of Particle technology at the University of Salerno is a 6 EC, that is a 60 hours course including lectures, as well as classroom and laboratory practical. The course dedicated to Master student in Chemical Engineering. It a compulsory course within the â€œinnovative process and nanotechnologies curriculumâ€ can be optioned by any mechanical or chemical engineering student at the master level.

The course includes basic particle size and shape characterization concepts, fundamentals on interparticle forces, fundamentals of powder mechanics and silo discharge, fundamentals on-fluid particle interactions and fluidization.

The main learning outcome is the ability of the student to understand the phenomena occurring in processes involving particulate solids and to carry out basic design operation on storage and fluidized bed units.

Copyright © American Institute of Chemical Engineers. All rights reserved.

507382 Professional Master of Engineering Degree in Particle Technology

Abstract:

Particles are ubiquitous in the process industries yet nearly absent in the undergraduate education of chemical engineers. This poster descibes a professional Master of Engineering degree program at University of Delaware (the MEPT) that closes that gap by exclusively focusing on particulate matter, its processing, and its incorporation into products. The program can be completed in a single year although we welcome already-employed students who wish to take it part time. Now in our fourth year, we will describe our program and our experiences including guiding distance learners successfully through the program.

In its 1-year realization, a core curriculum of 18 credits is augmented by 6 credits of electives which the student may use to tailor the program to their specific interests. The program is capped by a 10-week, 6-credit summer internship. The 12 core credits taken in the fall teach the fundamentals of particle transport, particle properties & characterization, particle formation and transformation kinetics and the mathematics of particulate systems. The 6 core credits taken in the spring teach the student to apply the fundamentals to both (a) equipment design (in a unit operations course) and (b) product and process design (in a product engineering course).

The presenters of this poster are the program directors, both with long industrial careers preceding this venture into education. In addition, we have developed a strong network of industrial partners who share a wealth of additional experience and expertise with our students both as adjunct professors and as internship mentors. We have successfully placed international students in internships at US companies via the curriculum practical training program (CPT) who have gone on to practice in industry after graduation under the optional practical training program (OPT).

Copyright © American Institute of Chemical Engineers. All rights reserved.

507780 Recent Developments in Particle Technology at the Universidad Tecnica Federico Santa Maria

Abstract:

The Department of Mechanical Engineering of the Universidad Técnica Federico Santa María formed in 2003 a Research Center for Transporting Granular Materials (CITRAM) to support educational activities and experimental research, as an extension for those ME students taking the regular course on Bulk Solids Handling and those developing a thesis in this area.

Over the years, many ME students have studied, investigated and modeled different topics including: pneumatic conveying of solid particles of different nature, size and density; identification of flow patterns developed in the pipeline, minimum conveying velocity and pressure drop; abrasive wear using a novel rubber-wheel tester; segregation of ores in stockpiles and silos; flow of granular materials in hoppers and chutes; heat transfer of granular materials; among others. The aim of this poster is to present and share the results of this work.

Copyright © American Institute of Chemical Engineers. All rights reserved.

508520 Using Perusall to Enhance Student Learning of Particle Technology at Graz University of Technology

Abstract:

Perusall is a digital platform that allows students to collaboratively annotate textbooks, papers, and other similar classroom reading assignments. In order to enhance student learning, Perusall was introduced to help teach Particle Technology at Graz University of Technology (TU Graz, Austria). At Graz University of Technology, two Particle Technology courses exist in sequence; the first course is focused on fundamentals topics, whereas the second course is focused more on applications and separations topics. Approximately 60-75 students are registered for each course. For both courses, Perusall was used to assign reading materials, which were primarily journal articles that complemented the course topics. Although these assignments were non-compulsory due to national requirements, students were asked to use Perusall to read and annotate the course readings. Student participation and input regarding this digital approach to learning was recorded. This poster will highlight how Perusall was implemented in both Particle Technology courses as a tool to enhance student learning.

Copyright © American Institute of Chemical Engineers. All rights reserved.

508524 A Graduate Course in Fluidization and Gas-Solid Flow Systems

Abstract:

The objective of this course is: a) to provide students with a fundamental understanding and design of Fluidization and Fluid- Particle flow systems, b) to provide students with an understanding of Computational Fluid Dynamics (CFD) and the kinetic theory approach to fluid-particle flows c) to provide students with the knowledge required to solve flow, heat and mass transfer problems in processes which involve particles and d) to provide students with sufficient knowledge to read and understand technical articles in areas related to fluid-particle processes. The course is designed for Master and PhD students in Chemical, Mechanical and aerospace engineering. The course includes: Fluidization concepts and regimes including bubbling, turbulent and circulating beds; Particle characterization, diagnostics and size reduction and agglomeration/granulation; Concept and design of Standpipes, non- mechanical valves, gas distributors, cyclones and hoppers; CFD approach to fluid-particle flow systems and fluidization including kinetic theory and dispersed phase modeling; Computational transport phenomena of fluid-particle systems including method of moments; and examples from energy, chemical and pharmaceutical industries. In addition the students will visit PSRI to see large-scale fluidized beds in operation.

The students are required to finish a team project. The teams consist of 3-4 students working on a project which will count as their final exam. The projects will primarily involve background research (about 80%) but some of the investigation (about 20%) should involve taking an additional step from what is currently known. This additional step can take the form of a research proposal, new theory, new insight based on current information, new calculations, new simulation and/or new data. The results of the projects will be communicated to the class and to the Course Instructors via a written report of 10-20 pages, and an oral presentation involving the professors and all of the group members.

Copyright © American Institute of Chemical Engineers. All rights reserved.

508742 Education of Fluid-Solid Multiphase Flow at Department of Mechanical Engineering, Osaka University

Abstract:

In mechanical engineering field, education curriculum in undergraduate course concentrates on the basic subjects of mechanical engineering such as mathematics, mechanics, material mechanics, fluid mechanics, thermodynamics, control engineering, design, etc. Concerning the course relating to powder technology, we have a class of "Fluid-Solid Multiphase Flow" in the curriculum of graduate school of engineering.

The contents of "fluid-solid two-phase flow" is as follows;

1 Introduction: examples of multiphase flow

2 Definitions and fundamental concepts

3 Particle response time and Stokes number

4 Fluid force acting on particle

5 Fluid force acting on particle in dense system

6 Particle-wall and particle-particle interaction

7 Heat and mass transfer

8 Continuous phase flow equations

9 Modeling of fluid-solid multiphase flow

10 Discrete particle modeling and simulation

11 Numerical simulation method for continuous phase flow

This course includes basic concepts of powder technology, dynamics of particles in fluid flow, derivation of basic equations for locally-volume-averaged fluid flow, discrete particle modeling including DEM (Discrete Element Method), and basics of numerical analysis and modeling.

Copyright © American Institute of Chemical Engineers. All rights reserved.

509420 Bulk Solids Handling Education at the KSU Bulk Solids Innovation Center

Abstract:

Kansas State University Bulk Solids Innovation Center (BSIC) was started in 2015 with university and industry partnership and is located in Salina, Kansas. This center is dedicated to bulk solids handling research, education and outreach and equipped with world class pneumatic conveying system (both dense and dilute) full scale test lab and other material testing equipment. The vision of this center is to become valued resources for companies who produce or handle bulk solids during their operations. Currently, this center offers short courses in various subject lines related to bulk solids handling such as storage and flow, pneumatic conveying systems, discrete element method, dust explosion and control and mixing and segregation. Duration for these short courses varies from two to three and half days and offered as live course with class room teaching and practical training. The fundamental course on bulk solids handling is offered as distance education online format. This center is also in the process of starting two year online (distance education) master degree program named professional master of technology specialized in bulk solids handling and one year graduate certificate program (online) in bulk solids handling.

Copyright © American Institute of Chemical Engineers. All rights reserved.

509079 Creating a Pathway to Sustainability: The Critical Contribution of Particle Technology Research

Abstract:

Maintaining adequate future supplies of energy and water requires the creation of a pathway to sustainable energy and water that is both multi-faceted and complex. An effective pathway aims to improve the reliability, security, and affordability of energy and water by: utilizing a least-cost strategy to reduce the negative impact of energy consumption on climate change and water availability; emphasizing the gradual decarbonization of the global energy system; increasing energy efficiency and conservation; and preserving natural resources and the environment.

A successful pathway to sustainability must include short-term, transitional, and long-term plans that must be launched simultaneously. This presentation will discuss the critical contribution of particle technology research to the successful implementation of these plans.

The role of circulating fluidized bed (CFB) research in carbon capture processes and particle pulverization as a recycling tool for polymeric and elastomeric materials as examples of the short-term plan will be presented.

As an example of the transitional plan, the impact of fluid-particle flow and computational fluid dynamics (CFD) research in the production of natural gas from unconsolidated hydrate reservoirs will be discussed.

The contribution of particle technology to the long-term plan, including the effect of rain droplets on wind turbine performance and the role of the fluid-particle mixture as a very effective heat transfer medium for concentrated solar power (CSP) systems will be presented.

Copyright © American Institute of Chemical Engineers. All rights reserved.

507204 Chemical Looping Combustion, Gasification and Reforming â€" Particle Technology Perspectives

Abstract:

The concept of chemical looping reactions has been widely applied in chemical industries. Fundamental research on chemical looping reactions has also been applied to energy systems. Fossil fuel chemical looping applications were used with the steam-iron process for coal from the late 1800s to early1900s and were demonstrated at a pilot scale for synthetic natural gas production with the IGT HYGAS Process and the CO_2 Acceptor Process in the 1960s and 1970s. There are presently no chemical looping processes using carbonaceous fuels in commercial operation. Key technical factors that determine commercial viability of the technology lie in the sustainability of the reactivity and recyclability of the metal oxide oxygen carriers and the ability of configuring the reactor assembly for optimal operation and control of the solids flow system. The chemical looping system is developed on the circulating fluidized bed platform. The successful deployment of this technology requires thorough knowledge of two interconnected fields, i.e., metal oxide reaction engineering and particle science and technology. With now CO_2 emission control of great concern and process conversion efficiency enhancement of great interest, activities on research and development of chemical looping technology have resurfaced.

Specifically, chemical looping technology is a manifestation of the interplay among such key elements of metal oxide reaction engineering and particle science and technology as particle synthesis, reactivity and mechanical properties, flow stability and contact mechanics, gas-solid reaction engineering and particulates system engineering. This presentation will describe the fundamental and applied features of modern chemical looping technology in the context of the circulating fluidized bed platform that utilizes fossil and other carbonaceous feedstock. It will discuss particle reaction engineering represented by reaction chemistry, ionic diffusion mechanisms, metal oxide synthesis and thermodynamics, reactor design, and system engineering along with energy conversion efficiency and economics of the chemical looping processes for full and partial or selective oxidation for electricity, fuel and chemicals production. The Ohio States University has developed a number of advanced chemical looping combustion, gasification and reforming processes which will be highlighted in this presentation. Potential for solar based chemical looping technology will also be discussed.

Copyright © American Institute of Chemical Engineers. All rights reserved.

507815 Dopant Modified Iron Based Oxygen Carriers for Chemical Looping Combustion and Gasification Applications (Invited)

Abstract:

Chemical looping technology is promising in high efficiency clean carbonaceous fuel conversion, including chemical looping gasification and chemical looping combustion. The reactions in chemical looping involve cyclic redox reactions with metal oxide as the oxygen carriers to provide lattice oxygen and regenerate in air. Iron-based oxygen carriers modified with both isovalent and aliovalent transition metal ions have been investigated in this work. Isovalent dopants can lower fuel conversion reaction energy barriers by providing active sites for fuel activation and conversion during redox reactions whereas aliovalent dopants can not only provide such active sites but also create Schottky defects to increase the ionic conductivity in the oxygen carriers. In this study, we examine the impact of both isovalent and aliovalent dopants on iron oxide oxygen carriers reactivity and conversion rates using methane reforming and CO combustion as examples. A screen of suitable transition metal ion dopants have been done by DFT calculation as well as experiments. A detailed characterization and reaction mechanism will be discussed. The understanding of the reaction mechanism provides insight into dopant modification for improving the reactivity and recyclability of oxygen carriers, which is essential in commercialization of the oxygen carriers for chemical looping gasification and combustion.

Copyright © American Institute of Chemical Engineers. All rights reserved.

508069 Porous Composites As Host Materials for Lithium-Sulfur-Batteries (Invited)

Abstract:

Scalability, efficiency and economic feasibility are important aspects for the fabrication of cathode materials for lithium-sulfur batteries. For the application of lithium-sulfur batteries in electric vehicles and stationary energy storage systems, different challenges need to be overcome to achieve sufficient energy and power density within weight and volume constraints. Further safety requirements and the reproducible fabrication of large active material quantities at low costs have also to be guaranteed for a wide range of applications [1]. Sulfur allows high theoretical capacities of 1675 mA h g^{-1} as cathode material in comparison to lithium, resulting in a theoretical energy density of around 2600 W h kg^{-1} for lithium-sulfur batteries [2-5]. However, using sulfur as active material leads to compositional and structural changes and large volumetric expansion up to 80 % during cycling, aspects which must undoubtedly be considered. Besides, reducing the formation of polysulfides as adverse byproducts and insulating Li_2S is still a challenge to be solved [1, 6, 7]. As a consequence of these insufficient cycle stabilities, low coulomb efficiencies [3] and high self-discharge rates [4, 8], a barrier has been built against commercialization [2, 7, 9-11]. Porous host materials with defined pore structures which are able to incorporate certain contents of sulfur have been postulated as a possible solution to compensate the drastic volumetric expansion during cycling. A widely-used method to incorporate sulfur within host materials is the melt-diffusion process.

As published in Zellmer and Titscher et al. [12], different aspects, such as the used host material and its specific properties, the parameters of the comminution process as well as the formulation strategy to prepare the electrodes were investigated to evaluate their influence on the electrochemical performance of the lithium-sulfur-battery. Concerning the host material, the content of sulfur, the pore volume, the particle size distribution, the crystallinity and the conductivity of the material were characterized. Subsequently, the host materials, prepared via hydrothermal carbonization or purchased, were used for the comminution process in different mills. As published in [13 -16] planetary ball mills can be used to incorporate sulfur into host materials. The process can be optimized by adjusting the process time and including thermal treatments. Based on the specific energy input, sulfur can be embedded within the material but at the same time pores within the material can be removed [13-16]. Experimental studies and DEM simulation were performed to characterize the operation of different mills with regard to the incorporation of sulfur and to manufacture homogeneous composites without the combustion of filled pores. A planetary ball mill was chosen in this study in comparison to the literature and additionally, a vibration mill was used to prepare the sulfur carbon composites under lower stress energies than the ones arisen during the process in the planetary ball mill. The vibration mill enables the preservation of the pores while assuring simultaneous homogenization of the product and shows considerable potential for the scale-up [17-19]. For the evaluation of the influence of the mechanical treatment in the vibration mill on the preparation of homogeneous composites, different carbon host materials with various pore size distributions were processed and characterized after each process step regarding the content of incorporated sulfur and the resulting particle size and pore size distribution. Different aspects were found in comparison to the investigations dealing with the planetary ball mill. On the one hand, smaller particle sizes and thus, lower contents of incorporated sulfur within the carbon host material were detected due to the higher specific energy within the planetary ball mill. On the other hand, more specific power is transferred into friction events, and thus heat can be dissipated leading to an improved incorporation of sulfur in the pore structure. This fact can also be discerned and determined via DEM simulations. Further on, only a homogenization and not an elimination of the composite (sulfur and host material) take place because of the lower frequency of events in normal direction [20]. Subsequently, electrodes were produced in order to characterize the material, understand the fabrication process and determine the electrochemical performance of the manufactured composites. To this aim, water was used as environment-friendly solvent and PEDOT:PSS as conductive binder to fabricate cathodes for such lithium sulfur batteries [6, 21]. The influence of the composite structure after the manufacturing within the different mills was also discerned via the electrochemical characterization. After analyzing the electrochemical performance, it was clearly observed that higher specific capacities at low cycle numbers were reached using the composites prepared via the planetary ball mill process. Nevertheless, a constant decrease of the specific capacity was also detected over the lifetime. In the case of the composites manufactured in the vibration mill, lower specific capacities for the first cycles were measured, but constant capacities over the whole lifetime were achieved. Additionally, significant influences of the different host materials on the electrochemical performance were also investigated.

Various particle sizes and thus, different formulation strategies and electrochemical performances result from the specific energy input based on the different comminution devices, namely vibration and planetary ball mill. In current studies, the focus is set on the usage of a continuous process (vertical stirred media mill) starting with the homogenization of different porous, conductive host materials combined with sulfur as active material. Furthermore, the addition of different solid and liquid electrolytes can be experimentally fulfilled in the mill. Besides, DEM simulations were coupled with the experiments in order to support the outcomes, compare previous investigations and give insight into the process.

References

[1] A. Manthiram, Y. Fu, Y.-S. Su, Accounts of Chemical Research 46(5) (2012) 1125-1134.

[2] X. Ji, K.T. Lee, L.F. Nazar, Nature Materials 8 (2009) 500-506.

[3] J. Gao, M.A. Lowe, Y. Kiya, H.D. Abruña, Journal of Physical Chemistry C 115 (2011) 25132-25137.

[4] B. Guo, T. Ben, Z. Bi, G.M. Veith, X.-G. Sun, S. Qiu, S. Dai, Chemical Communication 49 (2013) 4905-4907.

[5] N. Jayaprakash, J. Shen, S.S. Moganty, A. Corona, L.A. Archer, Angewandte Chemie 123 (2011) 6026-6030.

[6] Y. Yang, G. Yu, J.J. Cha, H. Wu, M. Vosgueritchian, Y. Yao, Z. Bao, Y. Cui, ACS Nano 5(11) (2011) 9187-9193.

[7] Y.V. Mikhaylik, J.R. Akridge, Journal of Electrochemical Society 151 (2004) A1969.

[8] V.S. Kolosnitsyn, E.V. Kuzmina, E.V. Karaseva, Journal of Power Sources 274 (2015) 203-210.

[9] L. Chen, L.L. Shaw, Journal of Power Sources 267 (2014) 770-783.

[10] S.S. Zhang,, Journal of Power Sources 231 (2013) 153-162.

[11] Y.M. Lee, N.S. Choi, J.H. Park, J.K. Park, Journal of Power Sources 119 (2003) 964-972.

[12] S. Zellmer, P. Titscher, E. Wienken, A. Kwade, G. Garnweitner, Energy Storage Materials (2017), 9, 70-77

[13] J. Xu, J. Shui, J. Wang, M. Wang, H.-K. Liu, S.X. Dou, I.-Y. Jeon, J.-M. Seo, J.-B. Baek, L. Dai, ACS Nano 8 (2014) 10920-10930.

[14] M.R. Kaiser, J. Wang, X. Liang, H.-K. Liu, S.-X. Dou, 9th International Conference on Lead-Acid Batteries â€“ LABAT 2014 279 (2015) 231-237.

[15] M. Nagao, A. Hayashi, M. Tatsumisago, T. Ichinose, T. Ozaki, Y. Togawa, S. Mori, Journal of Power Sources 274 (2015) 471-476.

[16] Y. Cao, X. Li, I.A. Aksay, J. Lemmon, Z. Nie, Z. Yang, J. Liu, Physical Chemistry Chemical Physics 13 (2011) 7660-7665.

[17] T. Yokoyama, K. Tamura, H. Usui, G. Jimbo, International Journal of Mineral Processing 44-45 (1996) 413-424

[18] T. Inoue, O. K., International Journal of Mineral Processing 44-45 (1996) 425-435.

[19] H. Huang, J. Pan, P.G. McCormick, Materials Science and Engineering A 232 (1997) 55-62.

[20] S. Zellmer, P. Titscher, C. F. Burmeister, A. Kwade, G. Garnweitner, (2017) Fraunhofer Verlag, Produktgestaltung in der Partikeltechnologie, Band 8, 65-77

[21] T. Stöcker, A. Köhler, R. Moos, Journal of Polymer Science Part B: Polymer Physics 50 (2012) 976-983.

Copyright © American Institute of Chemical Engineers. All rights reserved.

509516 Supported Nickel and Iron Oxide Nanoparticles for Catalytic Asphaltene Decomposition Under an Air/Vapor Atmosphere for Enhanced Oil Recovery (Invited)

Abstract:

The production of heavy and extra-heavy crude oils has increased recently due to the current energy demand and the depletion of conventional resources. However, these type of crude oils presents different problems related to its production, transporting and refining. Accordingly, several in-situ techniques have been developed for enhancing HO and EHO recovery and improve their properties. In this sense, nanoparticles have proven to work well in improving the physical and chemical properties of this type of oils, such as viscosity, API gravity, and content to heavy components such as asphaltenes.

This work aims to synthesize, characterize and evaluate for the first time supported nickel and iron oxide nanoparticles for catalytic asphaltene decomposition under an air/vapor atmosphere. Silica nanoparticles of 11 nm in size were synthesized by the sol-gel method and were characterized by size, SEM-EDX, surface area, FTIR, point of zero charge and thermogravimetric analyses (TGA). Metal oxide nanoparticles were synthesized over Silica surface following the incipient wetness technique using commercial reagents and solutions from laterite leaching. The extraction of Nickel-Iron consisted of a laterite leaching process by the use of solutions of sulfuric acid at various concentrations and the use of temperature. Which allowed as a final product a solution with these two components that were quantified using absorption spectrometry atomic, which yielded a majority of iron and its respective proportion in nickel according to the acid used. The particle size of nickel oxide and Iron oxide nanoparticles was estimated in 2 and 6 nm with 5.0 and 32.6% of dispersion over the support surface. Batch adsorption experiments performed for obtaining asphaltene adsorption isotherms for the support and the functionalized nanoparticles. It was observed that functionalized nanoparticles had higher adsorption capacity than the support. Further, TGA-FTIR experiments were performed for the asphaltenes in the absence and presence of the different nanoparticles evaluated.

A fixed asphaltene loading of 0.2 m^2/g was employed. Experiments conducted for an air flow of 100 mL/min up to 800°C. $H_2O_{(g)}$ was introduced to the system at a flow rate of 6.3 mL/min using a saturator filled with distilled water. Results demonstrated the enhancing of asphaltenes decomposition by the action of the nanoparticles due to the reduction in the decomposition temperature of the asphaltenes up to 200°C in comparison with the system in the absence of nanoparticles. This work opens a wider landscape about the use of functionalized nanoparticles for improving the efficiency thermal enhanced oil recovery processes in the presence of air and vapor.

Copyright © American Institute of Chemical Engineers. All rights reserved.

514426 Carbon Capture, Utilization and Storage (CCUS) via Innovative Mineralization Pathways

Abstract:

Ever increasing anthropogenic CO2 emission is one of the most urgent and difficult challenges faced by the global society. A wide range of carbon capture and storage technologies have been developed but the unprecedented scale of industrial CO2 emission limits the number of viable options. Ocean and alkaline rocks are two large natural sinks for CO2 but their uptake rates are currently not sufficient to keep up with increased CO2 emissions. Thus, the acceleration pathways of mineral carbonation have been investigated to capture and fix CO2 into thermodynamically stable solid carbonates. In-situ carbon mineralization in geologic formations has been suggested to lower the long-term monitoring costs, whereas ex-situ mineral carbonation is particularly interesting since they can use both alkaline rocks and alkaline industrial wastes. Moreover, ex-situ carbonation allows the formation of value-added products and their use can further offset the greenhouse gas emissions. In this study, Mg- and Ca-bearing minerals (e.g., serpentine) and industrial wastes (e.g., steel slags) have been engineered and activated to effectively leach out alkaline metals and to form carbonates with controlled chemical and physical properties. The use of Mg- and Si-targeting ligands as well as chemical or P_{CO2} (partial pressure of CO_2) swing and internal grinding have been investigated in terms of the extent of mineral dissolution and carbonation. First, Mg and other components are leached from minerals and industrial wastes via rapid ion exchange with proton, but Mg extraction is eventually limited by the formation of Si-rich passivation layer induced by non-stoichiometric dissolution (Mg and Si) and re-precipitation of amorphous silica on the reactive mineral surface. In order to extract additional Mg from the solid residue, internal grinding and chemical ligand systems are studied. Mechanical abrasion of the passivation layer significantly improved the overall leaching of Mg, particularly when this method is combined with ligands. A modified core-shell model has been developed and the simulation results agree with the experimental data and well predicts the regime transition from kinetically to mass transfer limited reaction regimes of mineral dissolution. Findings from this study suggests that the dissolution and carbonation of alkaline minerals and industrial wastes can be significantly accelerated to capture and store CO2 and mineralized carbon can also create unique large-scale utilization options for anthropogenic CO2 such as construction materials.

Copyright © American Institute of Chemical Engineers. All rights reserved.

PARTICLE SIZE CHARACTERIZATION IN MINERAL CARBONATION FOR UNDERSTANDING REACTION FUNDAMENTALS

Dr. Rafael M. Santos, University of Guelph, Guelph, ON, Canada

Introduction

Mineral carbonation is the conversion of carbon dioxide, in gas form or dissolved in water, to solid carbonates. These may include calcium carbonates, magnesium carbonates, and a variety of other alkaline earth metal carbonates. Alkaline earth metals can be derived from natural minerals, waste residues, or even brines. When performed in a reactor, mineral carbonation leads to the formation of a powder that can be composed of a carbonate alone (e.g. high purity precipitated calcium carbonate), or a mixture of one or more carbonates with other solid phases, such as silica and silicates. Materials characterization thus plays an important role in assessing the potential to use these carbonates in commercial applications (e.g. filler in paper, aggregate in concrete). Materials characterization also aids in understanding fundamental phenomena about the reactions, such as the behavior of the reagents, the formation of products and by-products, and the reaction rate and conversion limitations. This paper highlights findings of fundamental nature that the author has made on topics related to mineral carbonation, and that were made possible by assessing particle size, particle size distribution and other morphological characteristics.

Laser diffraction analysis (LDA) is a technique that is useful to follow particle size growth or reduction. Thus, it can help elucidate effects of carbonation, such as formation of precipitated layer or alteration of microstructure, and of sonication (a technique to intensify the carbonation reaction), such as particle attrition and fragmentation. Of the three average particle diameters commonly obtained by laser diffraction (D50, D[4,3] and D[3,2]), the Sauter mean diameter (D[3,2]), being surface area sensitive, is the most important when it comes to the susceptibility towards mineral carbonation of powdery materials, since mineral carbonation reactivity is proportional to exposed surface area. This mean diameter is also found to be the most sensitive to sonication effects, as it best detects the formation of micron- to sub-micron sized fragments. The volume-moment mean diameter (D[4,3]), on the other hand, is useful in indicating the degree of erosion of larger particles. The volume-based particle size distributions and the average particle diameters presented in this paper have been determined by wet LDA (Malvern Mastersizer S) in sonicated deionized water, with a detection range of 0.06–878.7 μm.

Results and Discussion

Ultrasound has been investigated as a way to promote particle breakage during slurry carbonation, and to remove the carbonated shell or depleted matrix layers that surround the unreacted particle core, thus reducing diffusion limitations and exposing unreacted material to the aqueous phase (Santos et al., 2013). Sonication (Hielscher UP200S, 24 kHz, 200 W) was shown to increase the conversion (94% vs. 86% over 25 minutes) and the process kinetics (0.65 vs. 0.59 g-$CaCO_3$/min) of calcium hydroxide carbonation, compared to mechanical stirring. Furthermore, particle size reduction of calcium hydroxide and carbonate powders is significantly greater than of steel slag particles over 30 minutes sonication (Table 1). This is likely a result of lower hardness, confirming the premise that sonication can clear the surface of unreacted mineral particles from the rate-limiting coverage of a carbonated shell.

Table 1. Particle size reduction by sonication, expressed as Sauter Mean Diameter (Santos and Van Gerven, 2011). Re-used with permission from John Wiley and Sons (4316570759368).

Powder	Original D[3,2] (μm)	Sonicated D[3,2] (μm)	Reduction (%)
$CaCO_3$ <200 μm	5.6	4.8	-14%
$CaCO_3$ 200-500 μm	25.7	6.4	-75%
$Ca(OH)_2$	1.8	0.9	-51%
CC slag <200 μm	2.6	2.5	-5%

Ultrasound-intensified mineral carbonation

LDA results have been particularly useful in assessing the performance of sonication as a means of process intensification. The effect of sonication on passivating layer removal from the surface of carbonated slag particles was elucidated by measurement of particle size distributions and electron microscopy imaging of particle morphology. Figure 1 shows volume-based particle size distributions of AOD and CC slag before carbonation (original) and after four hours of carbonation with (US) and without (stirred) sonication. For both slags, similar shifts in particle size distributions take place. With stirring, the particle size distributions shift to larger particles; in fact, not only does it appear that large (10-100 μm) particles grow, it also appears that the amount of particles smaller than 10 μm decreases significantly, signifying that agglomeration/joining of carbonated particles takes place. In the case of sonication, the effect is opposite: particle size distributions shift to smaller sizes, with clearly distinguishable creation of particles in the 0.2-1 μm and 2-10 μm ranges, and reduction in the fraction greater than 10 μm. It can be theorized that the smallest particles (0.2-1 μm) are formed by flaking of passivating layers, and the second mode (2-10 μm) is formed by fragmentation of slag particles and/or are the eroded remains of once larger slag particles.

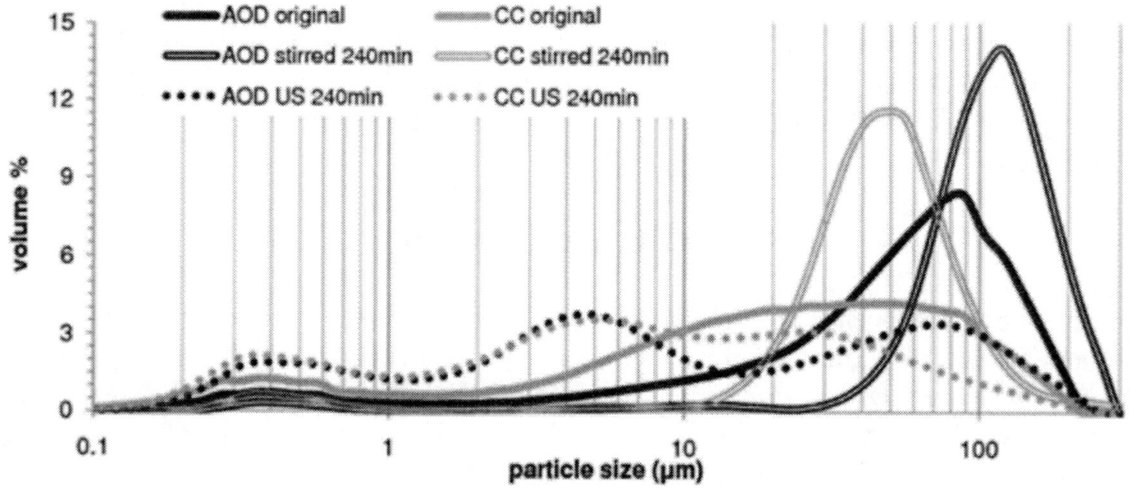

Figure 1. Particle size distributions of original and carbonated (240 min) powders of AOD and CC slags (Santos et al., 2013). Re-used with permission from Elsevier (4316570293874).

Figure 2 presents the average particles sizes of slag samples before and after carbonation as a function of reaction time. In the case of AOD slag under stirred carbonation, average particle sizes significantly increase after already 30 minutes of carbonation, but only D[3,2] increases further with reaction time. This is in line with the fact that AOD carbonation with stirring reaches nearly maximum conversion in the first 30 minutes; subsequent increase in D[3,2] suggests aggregation of smaller particles with the larger ones. With sonication, however, average particle sizes progressively decrease as a function of time, indicating the continuing effect of ultrasound abrasion/milling. In the case of CC slag carbonation, average particle sizes progressively decrease with sonication, and increase without. It is interesting to note that CC particle size already clearly reduces after only 30 minutes of sonicated carbonation, when in the case of non-reacting sonication the particle size did not change (Table 1). This is suggestive that sonication removes not only the precipitated calcium carbonate layer, but also the depleted silica layer that once constituted the original slag particle material, and that this depleted silica layer is weaker than the original silicate material.

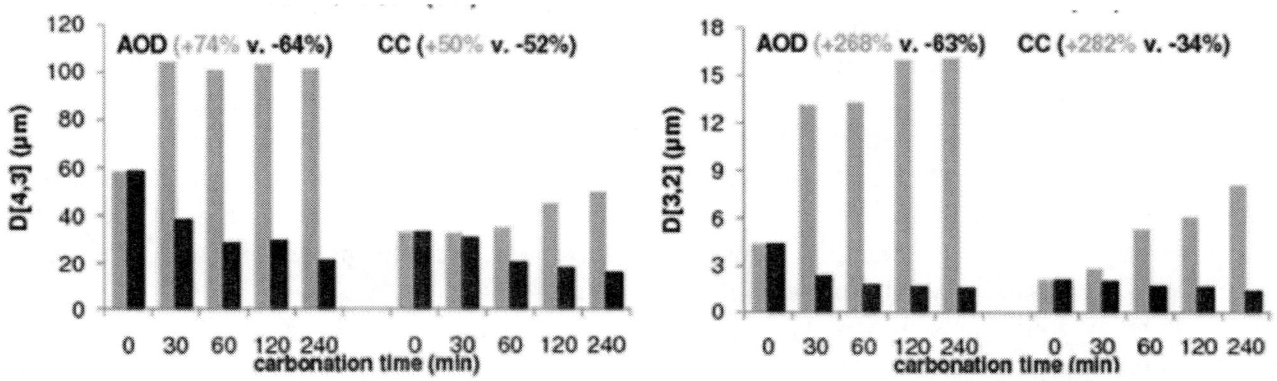

Figure 2. Average particle sizes of AOD and CC slags as a function of carbonation time with stirred and sonicated carbonation; percentage values indicate increase or decrease figures over four hours carbonation (Santos et al., 2013). Re-used with permission from Elsevier (4316570293874).

This has been further substantiated by comparing Scanning Electron Microscopy (SEM, Philips XL30) images to the LDA results. The morphology of CC slag particles prior and subsequent to carbonation is illustrated in Figure 3. As already indicated by particle size distributions, a clear distinction is seen between stirred and sonicated carbonation, and in relation to the original uncarbonated sample. In the case of stirred carbonation, the absence of particles smaller than 20 μm is confirmed, and the envelopment of slag particles in thick calcium carbonate layers, composed of rhombohedral calcite crystals, is validated. On the other hand, the sample subjected to sonicated carbonation exhibits distinguishably different morphology, being composed mainly of much smaller particles. Two modes are visible, in agreement with LDA results: flaky sub-micron particles, and rounded particles roughly 2-10 μm in size. It would appear that these rounded particles are either remnants of larger particles that have been significantly eroded over time, or particles whose rugged corners have been polished by sonication. These results substantiate the supposition that ultrasound is capable of removal of passivating layers leading to exposure of unreacted particle core to the reacting solution, overcoming the rate/conversion limiting shrinking core phenomenon and contributing to the intensification of the mineral carbonation reaction. It is also worth noting that, while SEM provides direct visualization of particles, it is a semi-quantitative method for assessing particle size distribution, while LDA is a quantitative method for doing so, and thus offers added insight not realizable by SEM.

Figure 3. Comparison of particle morphology of original (A), stirred carbonated (B), and sonicated carbonated (C) CC slag powders (Santos et al., 2013). Re-used with permission from Elsevier (4316570293874).

Synthesis of pure aragonite by sonochemical mineral carbonation

LDA also provides a window into the mechanism that controls the particle size of sonochemically synthesized aragonite precipitates (aragonite is the high-temperature polymorph of calcium carbonate). The formation of aragonite under sonication is attributable to two mechanisms: (i) imploding cavities generate localized regions of high temperature that can lead to the nucleation of aragonite seeds, given aragonite formation is promoted at higher temperatures; and (ii) the nucleation rate may be enhanced by sonication, thus maintaining stochastically preferable aragonite formation over calcite even at low crystal growth bulk temperatures.

Pure aragonite precipitates were synthesized in this study (Santos et al., 2012) at a range of temperatures spanning from 24 °C to 70 °C. As was observed by SEM analysis, particle morphology changes as a function of temperature and so does particle size. Particle size and particle size distribution are important parameters of powder materials for certain industrial applications such as paper coating or polymer filler; however, these data are not typically reported in literature for novel calcium carbonate morphologies, making it impractical to assess their suitability for industrial use. Table 2 lists average particle sizes by volume, expressed as D50 (mean volume diameter), D[3,2] (surface area moment mean diameter) and D[4,3] (volume moment mean diameter), and Figure 4 shows the volume-based particle size distributions.

Table 2. Average particle sizes of pure aragonite synthesized with ultrasound at different temperatures (Santos et al., 2012). Re-used with permission from Elsevier (4316570496158).

Temperature °C	D50 µm	D[3,2] µm	D[4,3] µm	Aragonite %
70 °C	4.15	1.31	6.48	99.5%
55 °C	1.33	0.78	1.87	98.7%
40 °C	0.98	0.65	1.27	98.6%
30 °C	0.85	0.69	1.23	99.3%
24 °C	0.84	0.79	1.09	98.6%
Fed-batch (30 °C)	1.74	0.87	2.22	96.7%

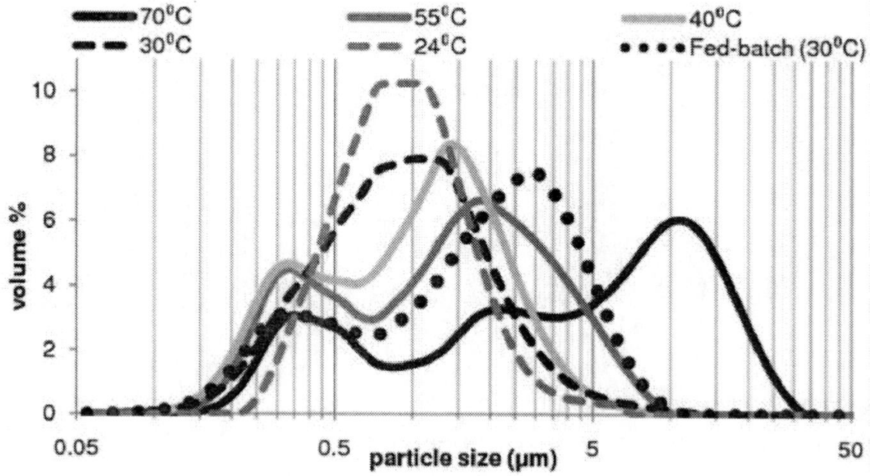

Figure 4. Particle size distributions of pure aragonite precipitates at varying temperatures (Santos et al., 2012). Re-used with permission from Elsevier (4316570496158).

All samples presented in Figure 4 were synthesized with ultrasound and contain greater than 98% aragonite, as indicated in Table 2. In general, lowering the synthesis temperature results in shifting of the distribution to the left (smaller sized); this is confirmed in Table 2 for the three listed size averages. The distributions of precipitates synthesized at higher temperatures have bimodal/trimodal characteristics, having a primary peak between 1 µm and 20 µm (and an intermediate peak at 2 µm for the 70 °C case) and a secondary peak below 1 µm. Larger particle modes are thought to represent needle-shaped particles, while the smaller modes may be fragmented particles, due to sonication, and/or particles that form near the end of the experiments when crystal growth becomes hindered by diminishing calcium concentration in solution. Samples from low temperature synthesis (24 °C and 30 °C) have unimodal distributions below 3 µm. The low temperatures, which may hinder crystal growth throughout the experiment, could be attributed to the smaller particle sizes. The fed-batch sample was found to have a bimodal distribution, likely comprised, as suggested by SEM imaging (Santos et al., 2012), of larger clustered crystals and smaller ultrasound-cleaved particles.

Self-regenerative additive for the mineral carbonation of calcium-rich materials

When $MgCl_2$ is used as an additive during slag slurry carbonation, LDA helps to provide an explanation as to why $MgCl_2$ addition is only beneficial in the presence of sonication. Particle size reduction can be linked to the formation of acicular aragonitic crystals. In the case of sonicated experiments with $MgCl_2$, the preferential formation of aragonite, is clearly evidenced in Figure 5a; however, due to extensive attrition of the particles caused by sonication, the crystal morphology was not clearly discernible by SEM imaging (Santos et al., 2014). Carbonate composition of Figure 5a was determined by Quantitative X-Ray Diffraction (QXRD, Philips PW1830 equipped with a graphite monochromator and a gas proportional detector, using Cu Kα radiation at 30 mA and 45 kV, mineral identification using Diffrac-Plus EVA (Bruker) and mineral quantification performed by Rietveld refinement technique using Topas Academic v4.1 (Coelho Software)).

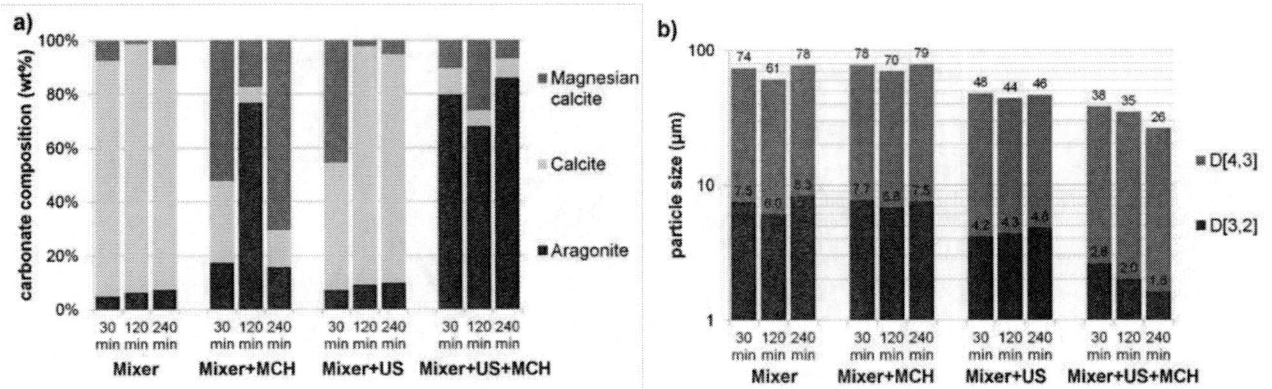

Figure 5. Carbonate composition (a) and mean particle diameters (b) of carbonated AOD slag using mixer-only, combination mixer/ultrasound, with and without MCH addition, as a function of reaction time; results with varying MCH dosages were averaged to facilitate plotting (Santos et al., 2014). Re-used with permission from Elsevier (4316570655781).

The formation of aragonite can be tied to the further reduction in particle size for experiments combining ultrasound with MCH (Figure 5b). The use of ultrasound, without $MgCl_2$, reduces the volume moment mean diameter (D[4,3]) of the carbonated slag from the range of 61–78 μm to a range of 44–48 μm, and the surface area moment mean diameter (D[3,2]) from 6.0–8.3 μm to 4.2–4.8 μm. The addition of $MgCl_2$ to sonicated tests further reduces the D[4,3] and D[3,2] to as low as 26 and 1.6 μm, respectively, after 240 min reaction time.

These reductions are significant, and can be explained by two mechanisms: (i) the erosion of large particles is indicated primarily by the D[4,3] value (as it is sensitive to particle volume changes), and (ii) the formation of micron- to sub-micron sized fragments is reflected on the D[3,2] value (which is sensitive to the formation of surface area). It thus appears that aragonitic crystals, due to their acicular morphology and arrangement, are more easily cleaved off the surface of the carbonated particles due to sonication, compared to the more closely-packed calcitic layers formed in the absence of $MgCl_2$. This occurs either through the induction of cavitation shock waves and micro-jetting on the particle surface, or through enhanced inter-particle collisions.

Carbonation Weathering Rate

Although several studies on metallurgical slag carbonation have been conducted, comparison of their reported reaction rate and conversion extent results has proven challenging. The main reason for this is that the slags used by the researchers have different particles sizes, and thus different specific surface areas. Mineral carbonation is a solid-state-diffusion-limited process under most conditions, so the different particle sizes affect the carbonation rate and maximal achievable conversion. In order to make direct comparisons among the results obtained from different studies, the Carbonation Weathering Rate (CWR) was conceptualized (Georgakopoulos et al, 2016). The rate is expressed in units of $\mu m \cdot min^{-1}$, and represents the weathering rate of the particle radius, from the original outer radius of the slag particle to the final radius of the unreacted core of the carbonated slag particle. It is calculated according to the following equation, where t_{carb} is the thickness of the carbonated shell of the particle, τ_{react} is the reaction time, r_x is the original average particle radius, and $C\%$ is the average particle conversion degree:

$$CWR = \frac{t_{carb}}{\tau_{react}} = \frac{r_x \cdot \left(1 - \sqrt[3]{1 - C\%/100\%}\right)}{\tau_{react}}$$

The CWR assumes that all reactive minerals carbonate at similar rates, that all particles are spherical, and does not account for the changing size of particles due to the accumulation of precipitated carbonates on the particle; that is, it only tracks the location of the reacted | unreacted interface. The second assumption is in principle not accurate, since the aspect ratio of iron- and steel-making slag particles is sometimes relatively high. However, laser diffraction is the typical technique used for determining average particle size, and this method does not distinguish particle shapes; therefore, the values used in the calculation of the CWR are already assumed to represent spherical particles. One limitation of the CWR is that it does not differentiate between weathering rate improvement due to particle size reduction and particle porosity enhancement. Thus, if a material is mechanically activated to improve carbonation, it will not be possible to distinguish between the two effects based on how the CWR responds.

By calculating the CWR from the carbonation conversions achieved by the studies that were reviewed by Georgakopoulos et al. (2016), Table 3 was created. Some studies refer to the size of the slag particles as a range instead of a specific average mean diameter. In those cases, an arithmetic average of the range was taken as the mean diameter to calculate the CWR.

Table 3. Summary of slag and process parameters from different carbonation studies and the resulting calculated Carbonation Weathering Rate (CWR) (Georgakopoulos et al., 2016). Re-used with permission from John Wiley and Sons (4316570882092).

	Type of Slag	Particle diameter (μm)	r_x (μm)	Conversion, C% (%)	t_{carb} (μm)	Reaction time, τ_{react} (min)	CWR (μm/min)
Huijgen et al. (2009)[41]	BOF	38	19	74	6.87	30	0.229
Chang et al. (2011a.)[42]	BOF	44	22	68	6.95	60	0.116
Van Zomeren (2012)[45]	BOF	2000-3300	1000-1650	4.67	16.95 - 27.97	3600	0.004 - 0.007
Santos et al. (2012)[48]	BOF	80	40	36	5.53	30	0.184
	BOF(wet)	125	62.5	20	4.48	1440	0.003
Chang et al. (2012)[44]	BOF	62	31	93.5	18.54	30	0.618
Chang et al. (2013)[43]	BOF	44	22	89.4	11.59	120	0.097
Polettini et al. (2015)[46]	BOF	63-100	31.5-50	53.6	7.11 - 11.29	240	0.030 - 0.047
Baciocchi et al. (2015)[47]	BOF(slurry)	150	75	40	11.74	1440	0.008
Tai et al. (2008)[33]	SS	63-90	31.5-45	65	9.30 - 13.29	60	0.155 - 0.221
Baciocchi et al. (2009)[49]	SS	105	52.5	27.2	5.26	480	0.011
Baciocchi et al. (2011)[36]	EAF(wet)	150	75	34.3	9.80	1440	0.007
	EAF(slurry)	150	75	25.4	7.00	240	0.029
Baciocchi et al. (2010)[38]	AOD	150	75	69.9	50.73	1440	0.017
Vandevelde (2010)[51]	AOD	38.7	19.35	32	2.33	8640	0.00027
	CC	40.7	20.35	45	3.67	8640	0.00043
Santos et al. (2011)[37]	AOD(mechanical)	60-230	30-115	30.5	3.43 - 13.14	240	0.0143 - 0.0547
	CC(mechanical)	60-230	30-115	61.6	8.20 - 31.41	240	0.0341 - 0.1309
	AOD(sonication)	60-230	30-115	48.5	5.95 - 22.82	240	0.025 - 0.095
	CC(sonication)	60-230	30-115	73.2	10.66 - 40.86	240	0.044 - 0.170
Van Bouwel (2012)[53]	AOD	46.1	23.05	63	6.50	60	0.108
	CC	39.3	19.65	76	7.44	60	0.124
Santos et al. (2013)[39]	AOD(wet)	46.1	23.05	24.2	2.03	8640	0.000235
	CC(wet)	39.3	19.65	37	2.81	8640	0.000325
Santos et al. (2013)[39]	AOD(slurry)	46.1	23.05	44	4.05	60	0.068
	CC(slurry)	39.3	19.65	57	4.82	60	0.080
Chang et al. (2011b.)[40]	BF	44	22	68.3	7.00	720	0.010
Cappai et al. (2015)[55]	Waelz	4000	2000	18.3	130.46	14400	0.009

As an overall observation, it is clearly shown that the highest carbonation conversions do not necessarily correspond to the highest CWR. It is worthwhile to point out the electric arc furnace (EAF) slag study conducted by Baciocchi et al. (2011). It is one of the few cases in the literature where wet (thin-film) carbonation of a particular slag results in higher conversion extent than when using the slurry route. However, the CWR calculated for the wet route is remarkably lower than that for the slurry route (0.010 μm/min for wet carbonation, versus 0.046 μm/min for slurry carbonation), indicating that the slurry route is the more time-efficient way for carbonation. In fact, the higher conversion that was achieved by the wet route could be attributed to the longer period of carbonation (6 days vs. 6 h). This comparison confirms the usefulness of the CWR calculation in delivering more accurate and insightful results as opposed to simple CO_2 uptake or conversion extent, since the CWR takes the reaction period into consideration, in addition to the particle size.

Conclusions

The purpose of this paper was to highlight how laser diffraction analysis (LDA) can be used in mineral carbonation studies to provide complimentary data and insight when used in combination with other techniques, such as scanning electron microscopy (SEM) and quantitative X-ray diffraction (QXRD). While LDA is a morphological assessment tool, it can aid in understanding a variety of chemical and mineralogical phenomena and mechanisms. The LDA technique was used in the reviewed examples as an *ex-situ* method, that is, to analyze materials before or after reaction. There are now particle size analysis tools that can analyze powders during reaction (*in-situ* and real-time), under elevated temperatures and pressures. These techniques utilize laser diffraction (Malvern), dynamic imaging (Sopat; Prozesstechnik), ultrasonic extinction (Sympatec), acoustic spectrometry (Dispersion Technology), and laser reflectance (Mettler Toledo). Even more insight than LDA can provide is thus possible, but mineral carbonation researchers are encouraged to at least incorporate LDA in their work.

References

1. Santos, Rafael M., François, Davy, Mertens, Gilles, Elsen, Jan and Tom Van Gerven (2013) "Ultrasound-intensified mineral carbonation". *Applied Thermal Engineering*, 57(1-2), pp. 154-163.
2. Santos, Rafael M. and Tom Van Gerven (2011). "Process intensification routes for mineral carbonation". *Greenhouse Gases: Science and Technology*, 1(4), pp. 287-293.
3. Santos, Rafael M., Ceulemans, Peter and Tom Van Gerven (2012). "Synthesis of pure aragonite by sonochemical mineral carbonation". *Chemical Engineering Research and Design*, 90(6), pp. 715-725.
4. Santos, Rafael M., Bodor, Marius, Dragomir, Paul N., Vraciu, Andrea G., Vlad, Maria and Tom Van Gerven (2014). "Magnesium chloride as a leaching and aragonite-promoting self-regenerative additive for the mineral carbonation of calcium-rich materials". *Minerals Engineering*, 59, pp. 71-81.
5. Georgakopoulos, Evangelos, Santos, Rafael M., Chiang, Yi Wai and Vasilije Manovic (2016). "Influence of process parameters on carbonation rate and conversion of steelmaking slags – Introduction of the 'carbonation weathering rate'". *Greenhouse Gases: Science and Technology*, 6(4), pp. 470-491.
6. Baciocchi, Renato, Costa, Giulia, Di Bartolomeo, Elisabetta, Polettini, Alessandra and Raffaella Pomi (2011). "Wet versus slurry carbonation of EAF steel slag". *Greenhouse Gases: Science and Technology*, 1(4), pp. 312-319.

508351 Viennagreen CO_2 Capture Pilot Plant Laboratory De-Risking, Design and Testing Objectives

Abstract:

Shell Global Solutions is a world leader in the design and development of regenerable liquid amine based flue gas scrubbing technologies. It offers solutions for the control of atmospheric pollution through the high efficiency post combustion capture of CO_2 from exhaust gases in a wide range of industrial applications.

In order to maintain its leadership position, Shell has joined, in 2015, the *ViennaGreenCO2* consortium which aims at developing the most cost competitive and environmentally friendly CO_2 capture technology in the world. This CO_2 capture technology uses a heat regenerable solid adsorbent in conjunction with 2 multi-stage fluidized beds where CO_2 adsorption and desorption occur.

The project has a duration of 4 years and is comprised of 3 main phases:

- **Phase 1**: Laboratory de-risking tests (e.g cold flow modelling, isotherm measurement) & process modelling.
- **Phase 2**: Design, Engineering & Construction of a 1 ton CO_2 per day pilot plant.
- **Phase 3**: Piloting campaign & Economic evaluation.

The pilot plant operation is scheduled to commence in late Q2 2018. Therefore, only Phase 1 and 2 along with the pilot testing program objectives will be presented to the audience.

Copyright © American Institute of Chemical Engineers. All rights reserved.

507640 A Novel Thin Film Deposition Method of Doped TiO2 Particles for Environmental Applications (Invited)

Abstract:

Nanoparticles are increasingly used for many applications targeting environment problems. The key challenge is its handling and performance. The latter is affected due to secondary processes such as aggregation and the former arise in transportation and safe disposal. This is solved by depositing the nano particle on a substrate, which are commonly termed as thin films. However the performance of thin films is limited to various physical factors, such as film thickness, configuration, material property etc which is often linked to the deposition method itself.

In this paper we present a novel deposition method named as chemical beam vapor deposition which allows the precise control of film growth rate and film configuration. The deposited film is charactirized and tested for its environmental application.

Copyright © American Institute of Chemical Engineers. All rights reserved.

507972 Synthesis of Surface Functionalized Adsorbents for Adsorption of Metal Ions and Organic Pollutants By Atomic Layer Deposition (Invited)

Abstract:

Different thicknesses of TiO_2 and Al_2O_3 films were applied on micron-sized silica gel particles and biochars by atomic layer deposition (ALD), respectively. TiO_2 coated silica gel particles were used as adsorbents for adsorption of metal and metalloid ions; their adsorption ability for a mixture of 19 trace elements of heavy metals and other toxic elements, including As(V), Se (IV), Be(II), Al(III), V(V), Cr(III), Mn(II), Co(II), Ni(II), Cu(II), Zn(II), Ba(II), Tl(I), Sb(III), Cd(II), Ag(I), Sr(II), Mo (VI), and Pb(II), from aqueous solutions was investigated. The removal efficiencies of As(V), Se(IV), V(V), Mo(VI), and Pb(II) were higher than 75% simultaneously in the mixed solution at pH 3 for 20 and 40 cycles of TiO_2 coated silica gel particles without heat treatment. In contrast, the uncoated silica gel particles did not adsorb any ions under this pH condition. Al_2O_3 deposited biochars were used for removal of organic pollutants and the adsorption capacities were evaluated by adsorption of methylene blue (MB). The data fit well with the Langmuir isotherm and the maximum adsorption capacities were found to be 26.8 and 35.0 mg/g at 25 °C for the uncoated biochar and 5 cycles Al_2O_3 (5Al) coated biochar, respectively. The adsorbed MB amount per square meter achieved 1.3 mg/m^2 onto the 5Al biochar, and it was twice the amount on the uncoated biochar. The improvement of adsorption capacity after Al_2O_3 ALD coating was due to its improved hydrophilia.

Copyright © American Institute of Chemical Engineers. All rights reserved.

Effect of Surface Modification on Filtration Performance of Gas-Liquid Coalescing Filters

Feng Chen, Zhongli Ji, Qiangqiang Qi*
Beijing Key Laboratory of Process Fluid Filtration and Separation, College of Mechanical and Transportation Engineering, China University of Petroleum, Beijing, PR China

Abstract

Great attention has been paid to the purification of liquid particles from exhaust streams by fibrous filters which are mostly wettable. However, it seems that non-wettable filters are more advantageous under certain conditions, with superior gas-liquid filtration performance. Hence, the requirement of modifying the surface properties of present wettable filters is increasing, because the fiber surface characteristics can not only alter shapes of liquid particles on the fiber surface, but also influence capture mechanisms of liquid particles carried by the airflow. Especially, the processes of coalescing, transporting and draining of liquid particles are closely related to the fiber surface properties.

The work aimed at experimentally investigating the effect of surface modification on separation of liquid particles by filters from air streams. Three different methods were applied to modify surface properties of glass microfiber filters, which were commonly in use in industries and were selected as the experimental control, being transferred from wettable to non-wettable. First method was using the silicon dioxide (SiO_2) nanoparticles, with a mean diameter of about 200 nm (greatly smaller than the fiber diameter), and they were physically grafted onto fiber surface. The second method was treating filters with fluoropolymer to reduce the surface energy of fibers. The third treating method combined the two strategies above, viz., nano coating and polymer fluorination. Therefore, the fibers possessing the best liquid-repellent performance were obtained. All filters were tested with an employment of a filtration performance experimental setup, aiming to evaluate the pressure drop, filtration efficiency and quality factor. The filtration efficiency was derived from both upstream and downstream particle number concentrations, simultaneously utilizing a scanning mobility particle sizer (SMPS) and an aerodynamic particle sizer (APS). The particle sizes taken in this study range from 0.05 to 20 microns.

It was illustrated in the results that the surface modification played an important role in the surface characteristics and filtration performance of fibrous filters. The three methods mentioned above were all able to change the wettability of filters in varying degrees. The average oil contact angles were 52°, 101°, and 121°, respectively, for these three different kinds of modified filters. It was found that the nanoparticles were uniformly distributed on the fiber surface by microscopic observation. Furthermore, a pronounced increase of equilibrium efficiency was shown from most modified filters, compared to the control. As expected, it was more difficult for the modified filters to be wetted by liquid, which resulted in a reduced equilibrium saturation, suggesting a longer operation lifetime for these filters. Meanwhile, the results also showed that all the modified filters presented a similar steady-state pressure drop as the control filter, indicating little effect of surface modification on the steady-state pressure drop of filters. On the whole, a filter, modified by combination of grafting nanoparticles and low-surface-energy polymer fluorination, had the overall best filtration performance in the steady state, with a highest quality factor of 0.69 kPa^{-1} for the droplets measured by the SMPS and of 2.02 kPa^{-1} for the droplets measured by the APS.

Keywords: Nanotechnology; Particle technology; Separations

1. Introduction

Gas-liquid coalescence filters are widely used in many industrial processes including long-distance natural gas transportation pipelines, compressor dry gas seal systems, metal and mechanical cutting processes, crankcase ventilation systems, compressed air purification, and so on, which are mainly used to remove liquid particles from air streams or oil mists [1]. These filters are commonly composed of fibrous materials, such as glass fibers, due to their high coalescence filtration performance [2]. It appears that most glass fibrous filters used in actual applications are wettable (or oleophilic), although non-wettable (or oleophobic) filters have obtained applications in some cases. Moreover, oleophobic filters are more advantageous in certain conditions due to less liquid retained in such filters resulting from low surface energy. Generally, the liquid retained in the filters will result in a reduced efficiency and increased pressure drop. Therefore, oleophobic filters may theoretically have a better filtration performance, which needs more investigation.

Research relevant with oleophobic filters mostly focus on three aspects, which are single fiber, sandwiched filter materials, and filter cartridge. At single fiber level, some work have been done on the interactive forces between fibers and droplets, droplet shapes on the fibers, and motions of droplets along fiber surface [3,4]. At filter material level, Kampa et al. [5,6] proposed a "jump-and-channel" model by experiments conducted with glass fiber filter media, and the effects of oil viscosity and filter properties on the pressure drop and liquid distribution were also analyzed. At filter cartridge level, the influence of oil loading rate and drainage layer on the filtration performance of oleophobic filters have been conducted by Chang et al. [7,8]. Additionally, two other work have been done on the effect of order of media wettability using a combined filter medium of wettable and nonwettable filters [9,10].

However, there are few works on the systematical analysis of gas-liquid filtration performance of filters with similar structures but different surface wettabilities. Best to our knowledge, there are only two relevant works. Letts et al. [11] investigated the effect of fiber surface wettability on pressure drop evolution and equilibrium classical filtration efficiency of three different filter materials. Their results showed that the high-surface-energy filter had a lower steady-state pressure drop due to less liquid hold-up, and initial pressure drop also differed among different filter materials. Another work was our previously published literature [12], in which we found that a superoleophobic filter that was modified using wet coating method presented a better filtration performance than untreated glass microfiber filter. Hence, surface modification is a useful method to prepare filters with unchanged structures but various surface energy, so that the effect of surface energy of filter medium can be analyzed. Nevertheless, there are many different methods to modify the filters, including addition of nanoparticles, fluorination, and combination of these two methods. Consequently, there is a growing need to analyze the effect of modification method on coalescence filtration performance of the filter. Also, this work may be of great potential meaning to the application and development of surface modification in the field of gas-liquid coalescence filtration.

The intent of this work is to analyze the effect of surface modification method on the filtration performance of glass fiber filters in terms of pressure drop, filtration efficiency, and quality factor, to study the liquid saturation of filters in the steady state, and to find an optimal treating method to obtain a filter with overall best filtration performance.

2. Experimental

2.1. Materials

The basic filter material used in this work was oleophilic glass fiber medium, of which the properties are listed in Table 1. Mean pore size was measured using a capillary flow pore size analyzer (CFP-1200AEF, Porous Materials, Inc.). Thickness was tested with a digital caliper. Grammage was obtained through weighing a filter piece with specific surface area using an electronic analytical balance (AL204-IC, Mettler Toledo). DEHS contact angle of filter material was measured using an optical tensiometer (Attension Theta, BiolinScientific).

Table 1. Properties of experimental glass-fiber filter materials

Filter material	Mean pore size (μm)	DEHS contact angle (°)	Thickness (mm)	Grammage (g/m²)	Packing density (-)
GF	4.53±0.02	44.0±4.1	0.56±0.01	102±3	0.070±0.002

All the filters were composed of two same layers, including four kinds of filter materials. Based on surface treating methods of non-treatment, nanoparticle-grafting, polymer fluorination, and combination of above two, the filters were termed filter GF, GF-P, GF-F, and GF-PF, respectively. Filter GF was regarded as the control filter. Following section is detailed modification procedures.

(1) Preparation of nanoparticles: 1.0 g silica nanoparticles were weighed and dispersed into 100 ml distilled water by stirring 3 h. (2) Preparation of coating solution: 5.0 g perfluoroalkyl acrylic copolymer was dissolved in 100 ml distilled water, then stirred by a stirring bar for 30 min to form 5 wt. % milky emulation solution. (3) Preparation of coating solution with nanoparticles: 1.0 g silica nanoparticles were added into 5.0 g perfluoroalkyl acrylic copolymer with magnetic stirring overnight. Fully dispersed the as-prepared slurry in 100 ml distilled water by stirring. (4) Coating treatment: The glass fiber filter was immersed in the coating solution for 10 min to ensure that all the fibers were completely wetted. Then the filter was taken out from the coating solution, and excess liquid was dripped in fume-hood. Finally, the filter was heated in an oven for 1 h.

2.2. Experimental set-up and procedure

Fig. 1 shows the schematic diagram of filtration performance experimental set-up. The oil used in this work was DEHS, with density of 912 kg/m³ and dynamic viscosity of 0.023 Pa s at 25℃. A six-jet aerosol generator (9306A, TSI) was used to produce the aerosol using compressed air purified by a HEPA (high efficiency particulate air) filter. The generated aerosol was mixed with clean air in a mixing chamber. The test filter whose efficient filtration area was 0.15 m in diameter was perpendicularly mounted in the filter chamber. The filtered gas was pumped using a vacuum pump after it was further purified by a HEPA filter. A mass flow controller was placed upstream the pump to control the face velocity during all experiments at 0.12 m/s. The captured liquid drained into a liquid collecting vessel under the force of gravity and air dragging force.

A scanning mobility particle sizer (SMPS, TSI) and aerodynamic particle sizer (APS, TSI) were used to simultaneously measure the upstream and downstream aerosol concentrations. By combination of these two instruments, droplets with diameters between 0.05 and 20 μm can be precisely analyzed. A pressure transmitter was used to continually measure and record the pressure drop of test filter. The measuring range and accuracy of each measuring equipment are listed in Table 2.

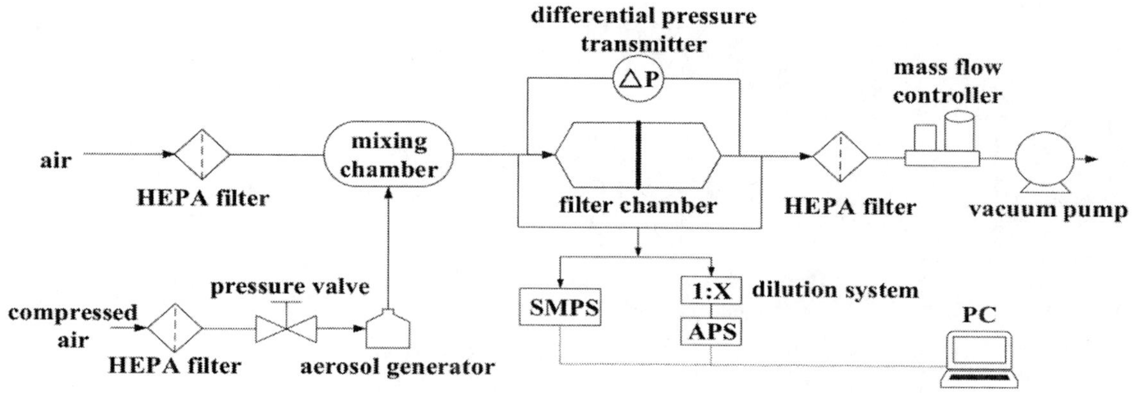

Fig. 1. Schematic diagram of filtration performance experimental set-up.

Table 2. Parameters of measuring equipment

Device	Model	Accuracy	Range
Mass flow controller	MCR 500 slpm	±0.4%	0–500 slpm
Differential pressure transmitter	EJX-110A	±0.04%	0–20 kPa
Electronic analytical balance	AL204-IC	0.1 mg	0–220 g
SMPS	TSI 3936	±4%	0.05–0.8 μm
APS	TSI 3321	±4%	0.8–20 μm

The filtration performance of filter is mainly evaluated in terms of pressure drop and filtration efficiency. To compare the overall filtration performance of all the filters, a quality factor (Q_F) was introduced [13] and defined by

$$Q_F = \frac{-\ln(C_{out}/C_{in})}{\Delta P} \qquad (1)$$

where C_{in} and C_{out} are the inlet and outlet droplets number concentrations, respectively, and ΔP is the pressure drop across the filter.

3. Results and discussion

3.1. Characterization

Fig. 2a-d shows the scanning electron microscope (SEM) images of filter GF, GF-P, GF-F, GF-PF. As shown, nanoparticles were evenly distributed on the fiber surface of filter GF-P and GF-PF. A magnified image of deposited nanoparticles on filter GF-PF is presented in Fig. 2e, and it seems that there existed an obvious agglomeration phenomena among nanoparticles, which needs a further work to be prevented and improved. It can be found that filter GF-F that was treated with polymer fluorination shows no perceptible difference from untreated sample. Fig. 2f shows the results of pore size of different filters. The pore size distribution and mean pore size were almost the same among all filters as shown in Fig. 2f, suggesting that there is little influence of fiber surface modification on the pore size characterization of filters.

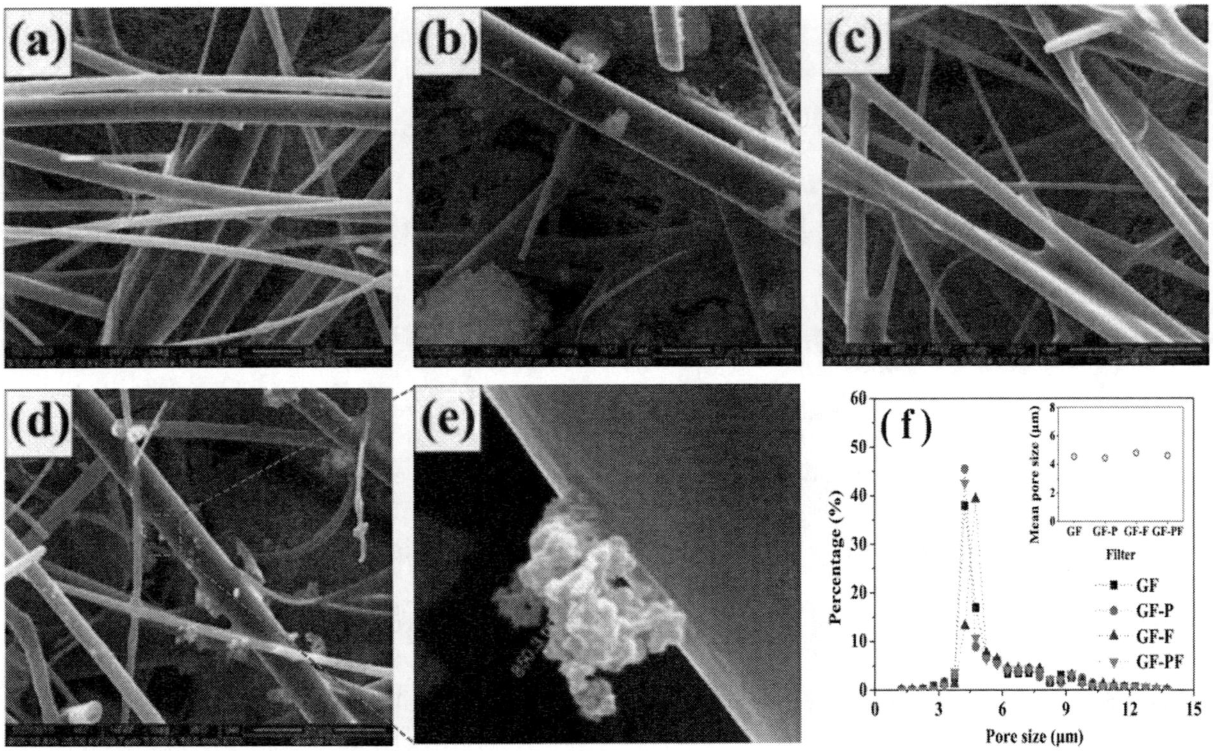

Fig. 2. SEM images of fibrous filter (a) GF, (b) GF-P, (c) GF-F, (d) GF-PF, and (e) magnified nanoparticles treated on the fiber of filter GF-PF; (f) pore size distribution of four filter materials.

Fig. 3 shows the DEHS contact angle and EDX result of different filters. As shown in Fig. 3a, the oil contact angle was improved for three modified filters when compared to the control filter. Filter GF-PF had a largest oil contact angle of 121.3°. Particularly, filter GF-F and GF-PF show an obvious oleophobic characteristic with average oil contact angles larger than 90°, indicating that the fiber wettability can be significantly changed through low-surface-energy fluoropolymer and addition of nanoparticles would further improve the oleophobic property of the filter. Fig. 3b shows that the increase of oil contact angle for filter GF-F and GF-PF was indeed related to the successful addition of element F on the fiber surface, due to relatively large amount of element F for these two filters while absent for other two filters. Besides, the contents of element Si and O appeared to greatly increase for filter GF-P and GF-PF after grafting nanoparticles on the fiber surface, as shown in Fig. 3b.

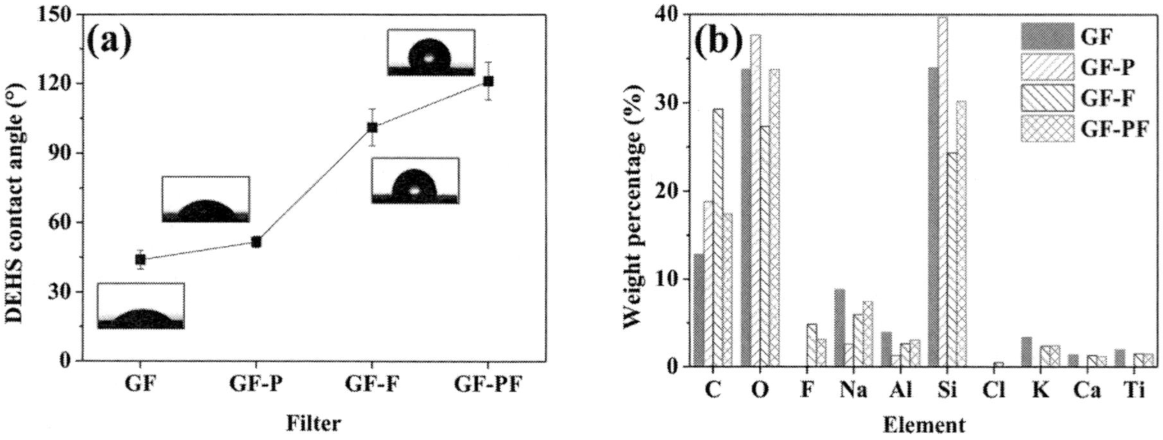

Fig. 3. Properties of different filter materials: (a) DEHS contact angle; (b) EDX.

3.2. Analysis of filtration performance

Fig. 4 shows the relationship between initial pressure drop and flow rate for filter GF-PF. As shown, initial pressure drop linearly increased with the air flow rate. Other filters illustrated a similar profile as this, suggesting that the fibrous materials were stably and evenly treated in some extents. By comparing the initial pressure drop under the experimental condition with face velocity remained at 0.12 m/s, as shown in Fig. 4, it was reasonable to conclude that there was little effect of surface modification method on the initial pressure drop of filter. This result was in good agreement with the previously mentioned result of similar pore size among different filters.

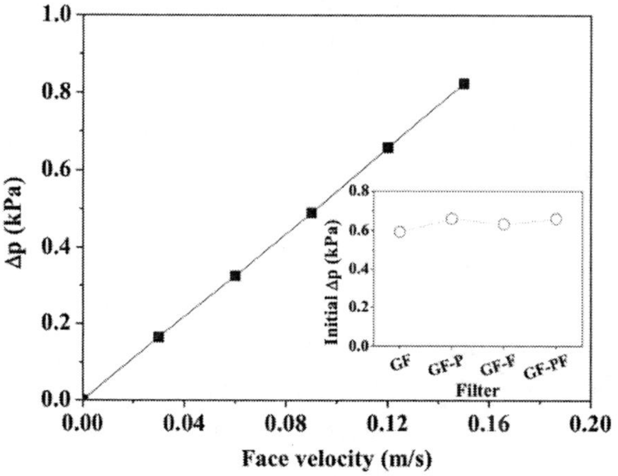

Fig. 4. Initial pressure drop of the filters.

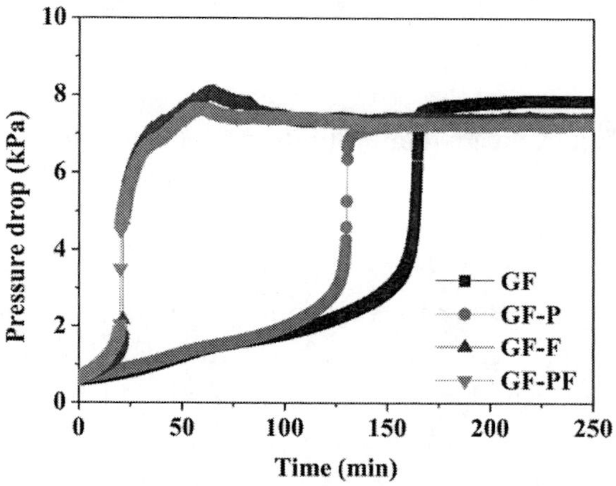

Fig. 5. Pressure drop evolution of the filters.

Fig. 5 shows the pressure drop evolution of the filters. Obviously, the steady-state pressure drop of three modified filters were almost the same, slightly lower than that of control filter. Because filter GF-F and GF-PF were both oleophobic, their pressure drop experienced a rapid increase after about 25 min from the experimental beginning, and then reached a steady state at approximately 100 min. By comparison, filter GF and GF-P presented a typical oleophilic trend, and the pressure drop increased slowly at early stage due to liquid channel transport, after which it appeared an earlier pressure drop jump for filter GF-P. Hence, it illustrates that a modification from oleophilic to oleophobic would fasten the evolution of pressure drop with time, which was more significant when the filter was relatively more oleophobic, while the steady-state pressure drop would basically not change.

Fig. 6 shows the steady-state filtration efficiency of the filters. As shown, filter GF-F and GF-PF had a significantly higher filtration efficiency for droplets in both particle size ranges when compared to the control filter. For filter GF-P, however, the efficiency was only slightly increased for droplets measured by the APS while decreased for those measured by the SMPS. It should be noted that filter GF-PF had the highest filtration efficiency for droplets measured by both particle sizers, suggesting that a combined modification of nanoparticles and fluoropolymer would maximize the filtration efficiency.

The steady-state quality factor was obtained after substituting filtration efficiency and pressure drop into Eq. (1), and the results are shown in Fig. 7. The quality factor of different filters illustrated a same fashion as the filtration efficiency shown in Fig. 6 due to their close pressure drop at steady state. Interestingly, filter GF-PF had the highest steady-state quality factor for droplets in both size ranges, meaning an overall best filtration performance. Therefore, it can be concluded that fiber surface modification by combination of nanoparticles and fluoropolymer, rather than only one of them, would most efficiently promote the filtration performance of the filter.

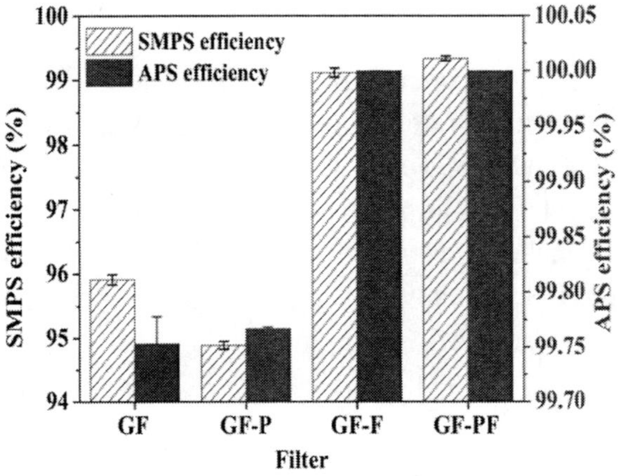

Fig. 6. Steady-state filtration efficiency of the filters.

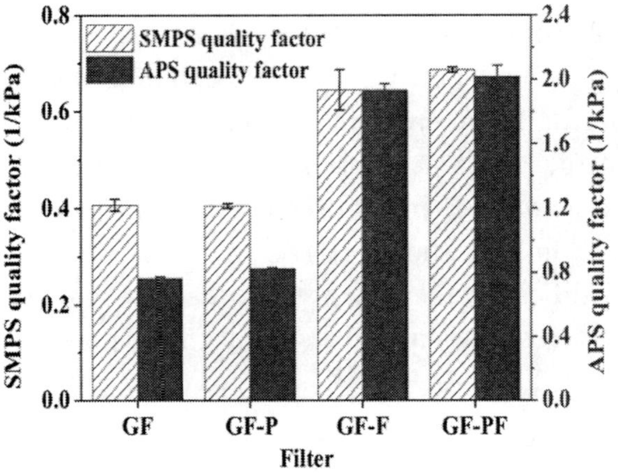

Fig. 7. Steady-state quality factor of the filters.

Fig. 8 shows the equilibrium saturation of the filters. For filter GF and GF-P, the saturation profile exhibited a typical oleophilic feature, where the second layer had an obviously higher saturation than the first layer. While for filter GF-F and GF-PF, the saturation of the second layer was largely lower than that of the first layer, on which a liquid film formed [5]. Note that the entire saturation of the last two filters was significantly lower than that of the former two filters. As a result, it is reasonable to suggest that the improvement of steady-state filtration efficiency of the filters was strongly related to the decrease of saturation, and the fiber surface modification was an efficient method to greatly reduce the saturation or liquid hold-up in the filter so that the filtration performance was promoted.

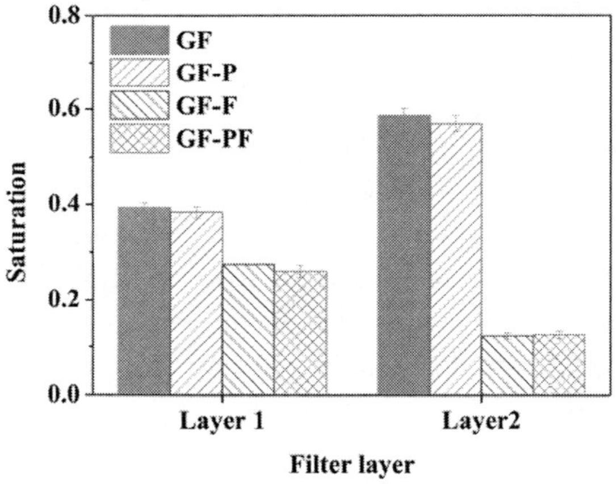

Fig. 8. Comparison of the saturation of different filters at steady state.

4. Conclusions

In this work the effect of surface modification on filtration performance of coalescing filters was experimentally conducted. The results showed that all surface modification methods used had little influence on the pore size and permeability of the filter. The oil wettability would not obviously change if the filter was only modified by nanoparticles, neither transferred from oleophilic to oleophobic. While modification by fluorination would effectively reduce the surface energy of the filter, thus increase the filtration performance. A combination of treatment by nanoparticles and fluoropolymer would produce a filter with highest oleophobicity, which was also of great advantage to the dispersibility and stability of nanoparticles on filter fibers. Moreover, the filter modified by this combined method exhibited an overall best filtration performance, due to significantly reduced liquid hold-up in the filter.

Acknowledgment

Authors acknowledge the financial support from National Natural Science Foundation of China under grant No. 51376196.

References

1. R. Mead-Hunter, A.J.C. King, B.J. Mullins, Aerosol-mist coalescing filters – A review, Sep. Purif. Technol. 133 (2014) 484-506.
2. G.M. Manzo, Y.R. Wu, G.G. Chase, A. Goux, Comparison of nonwoven glass and stainless steel microfiber media in aerosol coalescence filtration, Sep. Purif. Technol. 162 (2016) 14-19.
3. B.J. Mullins, R.D. Braddock, I.E. Agranovski, R.A. Cropp, R.A. O'Leary, Observation and modelling of clamshell droplets on vertical fibres subjected to gravitational and drag forces, J. Colloid Interface Sci. 284 (2005) 245-254.
4. R. Mead-Hunter, T. Bergen, T. Becker, R.A. O'Leary, G. Kasper, B.J. Mullins, Sliding/rolling phobic droplets along a fiber: measurement of interfacial forces, Langmuir 28 (2012) 3483-3488.
5. D. Kampa, S. Wurster, J. Buzengeiger, J. Meyer, G. Kasper, Pressure drop and liquid transport through coalescence filter media used for oil mist filtration, Int. J. Multiph. Flow 58 (2014) 313-324.
6. D. Kampa, S. Wurster, J. Meyer, G. Kasper, Validation of a new phenomenological " jump-and-channel " model for the wet pressure drop of oil mist filters, Chem. Eng. Sci. 122 (2015) 150-160.
7. C. Chang, Z.L. Ji, J.L. Liu, The effect of a drainage layer on saturation and liquid distribution of oleophobic coalescence filters, Sep. Purif. Technol. 194 (2018) 355-361.
8. C. Chang, Z.L. Ji, J.L. Liu, Pressure drop and saturation of nonwettable coalescing filters at different loading rates, AIChE J. 64 (2018) 180-185.
9. I.E. Agranovski, T. Myojo, R.D. Braddock, D. Jarvis, Combined wettable/non-wettable filter for mist purification, Chem. Eng. Technol. 24 (2001) 287-292.
10. B.J. Mullins, R. Mead-Hunter, R.N. Pitta, G. Kasper, W. Heikamp, Comparative performance of philic and phobic oil-mist filters, AIChE J. 60 (2014) 2976-2984.
11. G.M. Letts, P.C. Raynor, R.L. Schumann, Selecting fiber materials to improve mist filters, J. Aerosol Sci. 34 (2003) 1481-1492.
12. X. Wei, F. Chen, H.X. Wang, H. Zhou, Z.L. Ji, T. Lin, Efficient removal of aerosol oil-mists using superoleophobic filters, J. Mater. Chem. A 6 (2018) 871-877.
13. R.C. Brown, Air Filtration: An Integrated Approach to the Theory and Applications of Fibrous Filters, Pergamon Press, New York, 1993.

509070 Advances in Protic Ionic Liquids for Biomass Conversion into Biofuels and Bioproducts

Abstract:

The transformation of biomass into biofuels and bioproducts is of great importance in terms of bolstering the US bioeconomy and meeting national GHG reduction goals. Previous work at the DOE Joint BioEnergy Institute (JBEI, www.jbei.org) has demonstrated the capability of specific ionic liquids (ILs), such as 1-ethyl-3-methylimidazolium acetate ([C2C1Im][OAc]) and cholinium lysinate ([Ch][Lys]), to be effective biomass pretreatment solvents. Using these ILs for an integrated biomass-to-biofuel configuration is still challenging due to a significant water-wash process that is required due to the high toxicity of [C2C1Im][OAc] and pH adjustment prior to saccharification for the highly basic [Ch][Lys]. In this work, we demonstrate, We have developed and demonstrated a process based on the use of a protic IL (PIL), ethanolamine acetate, without pH adjustments, water-wash and solidâ€“liquid separations. After pretreatment, the whole slurry is directly used for simultaneous saccharification and fermentation (SSF) with commercial enzyme cocktails and wild type yeast strains, generating 70% of the theoretical ethanol yield (based on switchgrass). The structureâ €“performance relationships of PILs in terms of lignin removal, net basicity, and pH value are systematically studied. A technoeconomic analysis (TEA) revealed that an integrated biorefinery concept based on this PIL process could potentially reduce the minimum ethanol selling price by more than 40% compared to scenarios that require pH adjustment prior to SSF. Improvement of the economic performance will be made by reducing the dilution and enzyme loading during SSF as identified by TEA. This effort demonstrates the impact of a biocompatible PIL in terms of process optimization and conversion efficiency, and opens up avenues for realizing an IL based efficiently integrated biomass conversion technology.

Copyright © American Institute of Chemical Engineers. All rights reserved.

507625 Synthesis and Design of Porous Nanospherical Particles for Vehicle Electrification

Abstract:

Shifting fossil energy to low-carbon clean energy is critical to prevent global warming, and increase sustainable economic growth. In transportation sector, the gains obtainable from advanced technologies applied to ICEs (internal combustion engines) such as downsizing and combustion control improvements cannot of themselves solve the issues of CO2 emissions. Electrification of the automobiles through proliferation of HEVs (hybrid electric vehicles), EVs (electric vehicles), and FCVs (fuel cell vehicles), coupled with development of sustainable sources of primary energy, will be essential in the future. In this talk, we will present the synthesis and design of porous nanospherical particles for electrification of the automobiles due to their large surface area and unique properties which arise from scaffold compositions. The first example will present a synthesis platform for mesoporous metallic alloy particles by combining the colloidal chemistry and aerosol technique. Electron tomography demonstrates that the resulting mesoporous particles are assembled from primary nanoparticles into three-dimensional (3D) networks. Such high-surface-area mesoporous particles exhibit excellent thermal stability and hold potential in catalysis of hydrogen storage. In the second example, we report a size-dependent synthesis method for hierarchically porous Si nanospheres (hp-SiNSs), which consist of a porous shell and a hollow core. On charge/ discharge cycling, the hp-SiNSs accommodate the volume change through reversible inward Li breathing with negligible particle-level outward expansion. Our mechanics analysis revealed that such inward expansion is enabled by the much stiffer lithiated layer than the unlithiated porous layer. LIBs assembled with the hp-SiNSs exhibit high capacity, high power and long cycle life, which is superior to the current commercial Si-based anode materials. The low-cost synthesis approach provides a new avenue for the rational design of hierarchically porous structures with unique materials properties.

Copyright © American Institute of Chemical Engineers. All rights reserved.

Particle Measurement in High-Temperature Gas based on Light Scattering

Li-feng LU[1,3], Xiao-lin WU [1,3], Zhong-li JI [1,2], Ming-xing WANG[1,3]

[1] Department of Chemical Engineering, China University of Petroleum, Beijing, P. R. China

[2] Department of Mechanical and Transportation Engineering, China University of Petroleum, Beijing, P. R. China

[3] Beijing Key Laboratory of Process Fluid Filtration and Separation, China University of Petroleum, Beijing, P. R. China

Abstract: In order to achieve the particle measurement at 650 °C, a single particle measurement method and device suitable for high-temperature gas were developed. The influence of gas temperature on optical sensor of optical particle counter (OPC) was investigated in detail based on geometric optics theory and heat transfer theory. The thermal expansion of aerosol tube along the airflow direction was caused by the heat conduction between the tube and the gas, resulting in the position of the light apertures to shift to obscure the partial light beam. The invar material can be adopted to eliminate the above mentioned disadvantages. According to the principle of gas state equation, the optical parameters (refractive index) of the gas at different temperatures change with the change of its state parameters. Based on the geometrical optics theory and the geometric mathematical relationships, the theoretical and experimental dynamic models of the optical measuring volume with the change of the gas temperature were established. The established models were then used to analyze the influence of gas temperature on the optical measuring volume. The results showed that the influence of the temperature on the optical system is negligible. Meanwhile, particle detection experiments were conducted at different temperature to achieve a good performance of OPC.

Keywords: High-temperature gas; Optical measuring volume; Dynamic model

1 Introduction

With the continuous improvement of environmental protection and the safety requirements for industrial production, the technology of high-temperature purification is drawing more and more attention, especially in the fields of clean coal power generation, coal gasification, petrochemical industry and metallurgy [1]. For example, nuclear power plants often produce a large number of low radioactive solid wastes, and the combustible wastes usually account for a considerable proportion. The incineration and pyrolysis are effective treatment techniques for the combustible waste, during which the temperature of flue gas is 600~800 °C. With the aim to purify the flue gas, the German Karlsruhe Nuclear Research Institute, the French nuclear power station and the Japanese nuclear power station have installed the silicon carbide ceramic filter tube, aluminum silicate fiber and alumina ceramic filter tube at gas temperatures above 650 °C, respectively, and investigated their properties properties [2-4]. These filters have a direct influence on the purification quality of gas and the particle concentration in the downstream. In an industrial cracker system, the particle concentration is required to reach 0.014 mg/m^3 after the filters; the Integrated Gasification Combined Cycle (IGCC) requires the particle size to be smaller than 1 um with a particle concentration of less than 2.2 mg/m^3 and in the range of 1~10 um with less than 5.8 mg/m^3, and to remove the dust size of larger than 10 um [5]. In a Shell Coal Gasification Process (SCGP), the particle concentration is required to be less than 20 mg/m^3 [6]. In order to detect the concentration and size of the particles in pipes in real time for calculating the separation efficiency of evaluating the purification performance of the filtration system, the particle detection technology is needed to be improved [7].

The widely-used particle detection devices in industry are mainly based on the principle of light scattering [8-10], and the particle size and concentration are obtained through an inversion of the scattering light intensity and particle size, which are suitable for the conditions of below 200 °C. The method of first cooling and then detecting can be adopted for the particle detection under higher-temperature conditions, but droplet condensation and sublimation will inevitably occur and result in an incorrect particle samples.

With the aim to achieve the particle detection at 650 °C, an optimization was conducted on the existing particle detector (OPC) in this paper. And based on geometric optics theory [11] and heat transfer theory, a detailed research was performed on the influence of gas temperature on OPC optical sensor, including the dynamic change of the aerosol tube and the optical measuring volume (OMV). Finally, the experiments were conducted on the on-line particles measurement of different temperatures.

2 Experimental setup and method

Figure 1 gives the experimental setup for detecting particles in high-temperature gas, mainly including the heating unit, optical detection unit and gas purification and processing unit.

The heating unit consists of the vibration particle feeder and the air heater. The gas brings the particle out of the feeder, and is heated up to 1000 °C by the heater, which is monitored quickly by a PID temperature control instrument with easy operation.

The optical detection unit is used to achieve the experimental measurement of the characteristics of the OPC optical system and the on-line detection of particles, including the optical sensor and the OPC control host. The optical sensor is shown in Figure 2, which is composed of an aerosol tube and optical system. Aiming to avoid the influence of high-temperature gas on optical system, the aerosol tube (shown in Figure 3) has been developed, and is in a distance of 15 mm to the optical lens. In addition, the insulation coating with lower thermal conductivity (-80-1700 °C) is adopted to treat the outer wall of the aerosol tube. A K-type micro thermocouple was bent and installed into the light aperture to monitor the gas temperature at the center of the aerosol tube in real time without affecting the airflow and optical paths. Figure 4 shows the optical system of the sensor, which is composed of an incident light path and an observed scattering light path. Both the paths are arranged in 90° with the same optical lens and elements [12-14], and the intersected area is the center of the aerosol tube to form an OMV [15]. The intersected area is enough small and allows particles to pass through it one by one [16, 17]. The OPC host consists of the light source system (wavelength: 300-700 nm) [18], the photoelectric detection system and the signal analysis and processing system. In the particle detection, particle is illuminated by the incident light when passing through the OMV, and the scattering light is collected and transmitted to the photoelectric detector to achieve the conversion of a pulse voltage signal from an optical signal. Finally, the tested particle size is calculated based on the calibration curve of particle size and voltage amplitude [14, 19].

The gas purification and processing unit is composed of a vacuum pump, a flow controller, a dust collector and a heat exchanger. The high-temperature gas form the aerosol tube is completely cooled during the heat exchanger, the flow rate of which is adjusted by the flow controller. In order to avoid the damage of particles on the pump, the dust is captured and collected by a filter in the dust collector.

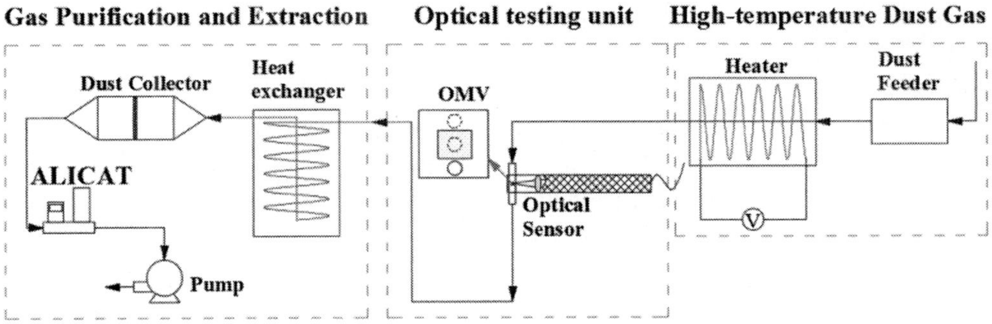

Figure 1. Experimental setup for detecting particles in high-temperature gas

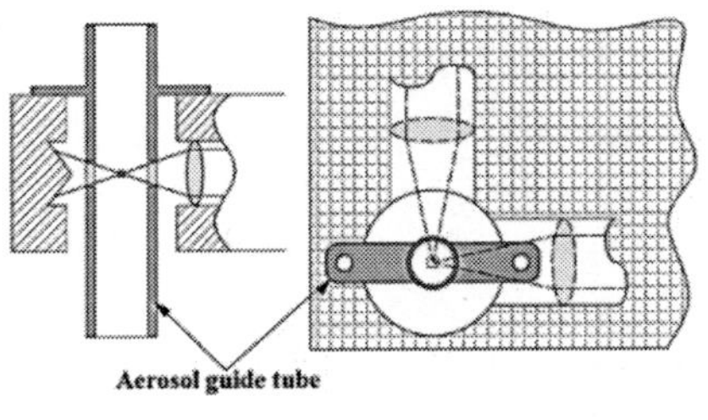

Figure 2. Optical sensor

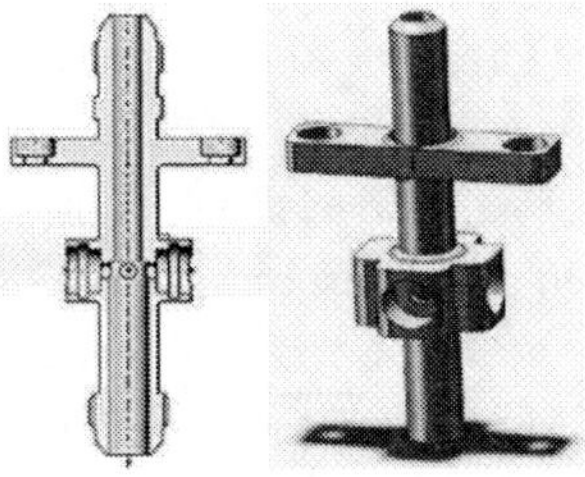

Figure 3. Aerosol tube for high-temperature gas

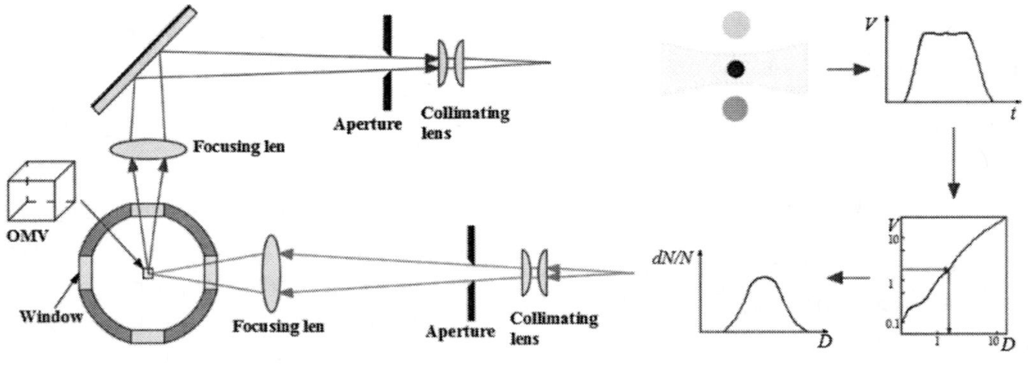

Figure 4. Optical system and principle of OPC

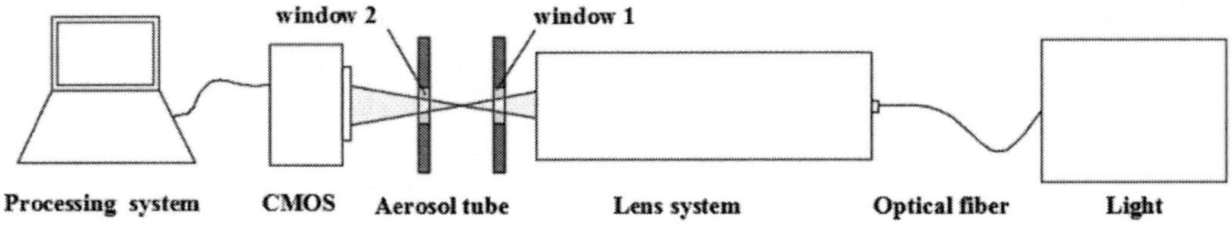

Figure 5. Evaluation and analysis device for light beam

3 Analysis and discussion

3.1 Influence of temperature on aerosol tube

The aerosol tube shown in Figure 3 has four apertures along the center axis, in which the glasses with a thickness of 2 mm were installed, as shown in Figure 4. When the gas flows in the tube, it can reduce the damage of high-temperature gas to the optical system elements. But the heat resistance of the tube has a direct influence on the imaging of the optical system at the center of the aerosol tube.

Figure 5 gives an evaluation and analysis device for light beam, including the light source system, the light path system, the aerosol tube and the light beam analyzer (CINOGY CMOS-1201, Germany), which can be equipped at the optical testing unit as shown Figure 1. For the gases at different temperatures, the light passes through the glasses in the aerosol tube and images on the CCD of CMOS-1201 to detect the spot of the light beam in real time.

After eliminating the background noise, the spot image at different temperatures can be acquired. When the aerosol tube is made of 304 stainless steel, the measurement results are shown in Figure 6. It can be found that the shape of the light spots is distorted and shows the imperfection of the imaging with an increase in gas temperature. That's due to thermal expansion along the airflow direction caused by the change of gas temperature, which results in the shift of light apertures position to obscure the partial light beam. A bad result must be acquired for smaller particle size and lower particle count efficiency because of the decreased incident light. Meanwhile, the shape of the OMV was changed to greatly reduce the performance of OPC detection by a failure of eliminating the error of edge detection in optical system.

With the aim to eliminate the above mentioned disadvantage, the invar material with a linear expansion coefficient of 1.6×10^{-6} m/°C was adopted for high-temperature experiment. The results are shown in Figure 7. At different temperatures, no deformation was found in the shape of light spots, indicating that the material can meet the testing requirement for high temperature up to 650 °C with a good heat resistance.

Table 1. Comparison of the application conditions of the two materials

Material	Light divergence angle /°	Gas temperature /°C
304 (Stainless steel)	<10	<200
4J36 (Invar)	<30	<700

The application conditions of the two materials are

summarized in Table 1. Compared with the stainless steel, the invar material is of great significance for the development of high-temperature detection device.

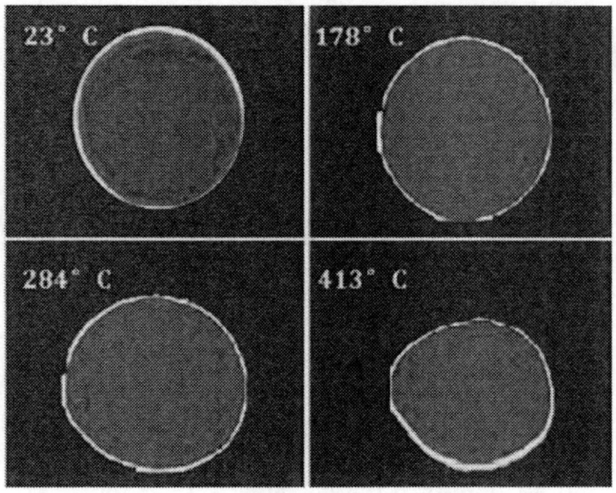

Figure 6. The spot images measured at four different temperatures of gas for 304 stainless steel.

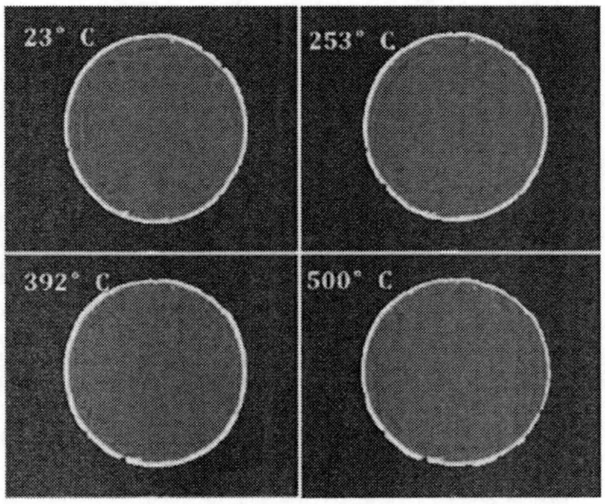

Figure 7. The spot images measured at four different temperature of gas for 4J36 (Invar).

3.2 Influence of temperature on optical system

In general the optical sensing system in the particle detection device is only applied at atmospheric temperature, but can also satisfy the temperature up to 150 °C. Under high-temperature condition, the gas volume is going to expand with the reduction in the gas density. The transmission characteristic of light beam could be changed, resulting from the decrease in the optical performance of the sensor. In order to investigate the influence of temperature on optical system, the specific causes are analyzed by theoretical method, and the extent of its influence is determined through experimental verification.

3.2.1 Theoretical method

By calculating the refractive index of the gas at different temperatures, the change of the OMV with the temperature can be obtained.

a) Calculation of gas parameters

The gas medium is air, so its refractive index can be calculated by the formula proposed by Rueger at the working temperature, given by:

$$n_{ph} = \frac{273.15}{1013.25}\cdot P\cdot N_{sph}\cdot\left(a_0+\frac{a_1}{T}+\frac{a_2}{T^2}+\frac{a_3}{T^3}+\frac{a_4}{T^4}\right)\times10^{-6}+1 \quad (1)$$

Where P is the gas pressure; a_0, a_1, a_2, a_3 and a_4 are the correlation coefficient and constant; N_{sph} is the gas refractive index at atmospheric temperature, given by:

$$N_{sph}=(n_{sph}-1)\times10^6=287.6155+\frac{1.62887}{\lambda^2}+\frac{0.01360}{\lambda^4} \quad (2)$$

The wavelength range is 400-700 nm. The gas refractive indexes at different temperatures in the range of 25-650 °C are calculated by Eq. 1 and 2. The results are shown in Figure 8. It can be found that the refractive index of air decreases with the increase in the gas temperature. And it can be inferred that the light divergence angle at high temperature is going to increase compared with that at atmospheric temperature. Consequently, the position of the beam focus is offset, leading to larger optical measuring volume, as shown in Figure 9.

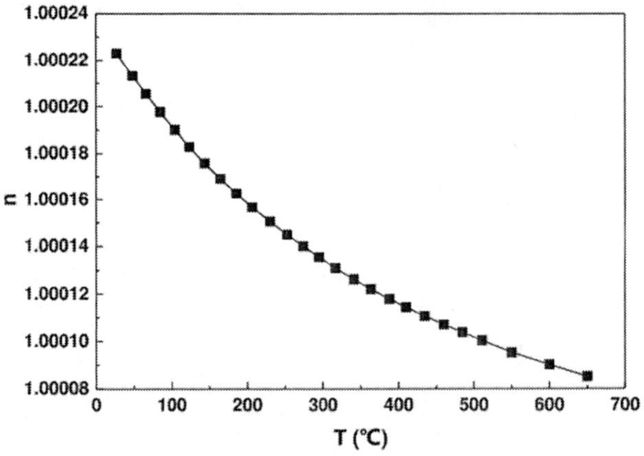

Figure 8. Gas refractive index at different temperature

In order to analyze the above unfavorable influence, a model of the OMV and gas temperature is needed to be built, mainly including the functions of the determination of the optical parameters and the calculation of the dynamic variable.

b) Calculation of optical parameters

On the basis of the Lorentz-Lorenz formula and actual gas state equation, the relationship between gas pressure and refractive index is given by:

$$\frac{n^2-1}{n_0^2-1}=\frac{\rho}{\rho_0}=\frac{PT_0Z_0}{P_0TZ} \tag{3}$$

Where T_0, P_0, ρ_0, Z_0 and n_0 are the gas temperature, pressure, density, compression factor and refractive index under atmospheric condition, respectively; T, P, ρ, Z and n are the gas temperature, pressure, density, compression factor and refractive index under high-temperature condition, respectively.

The light divergence angle is calculated by the refraction law shown below:

$$n_0' \sin\alpha = n \sin\beta \tag{4}$$

Where n_0' and α are the air refractive index and light divergence angle under atmospheric condition; β is the light divergence angle under high-temperature condition.

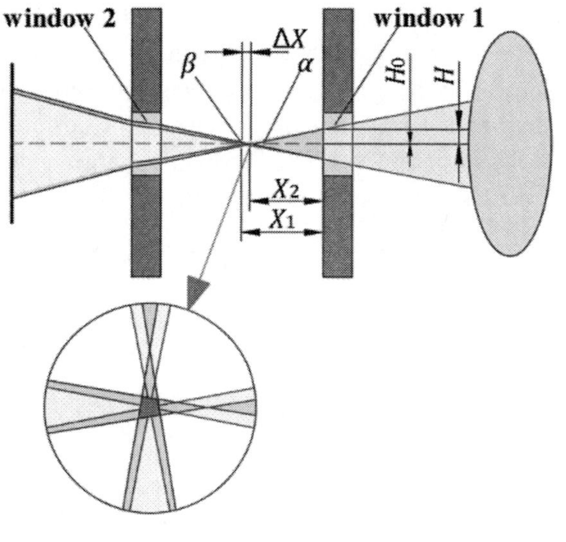

Figure 9. Schematic of the enlargement of the OMV

c) Calculation of dynamic variable

Based on the geometric optical theory and the mathematic relationship, the distance between the light focus position and the glass window 1 is triangular with the equivalent spot radius of the inner side of the glass window 1. According to X_1 and X_2 (the distance between the light focus position and the inner side of the glass window 1 under atmospheric condition and high-temperature condition), the theoretical model of the OMV with gas temperature was built. X_1 and X_2 are calculated by:

$$X1=\frac{H-H_0}{\tan\alpha} \tag{5}$$

$$X2=\frac{H-H_0}{\tan\beta} \tag{6}$$

The change of the light focus position (ΔX=$X1$-$X2$) can be calculated by Eq. 5 and Eq. 6.

The change of the equivalent spot diameter of the OMV can be calculated based on the triangle relation, as shown below:

$$\Delta D = 2\Delta X \tan\beta \tag{7}$$

Eq. 6 and Eq. 7 are the formulas of the theoretical dynamic model of the OMV with the change of the gas pressure.

Using the model, the optical parameters at different gas temperatures in the range of 25 ~ 700°C were calculated with three light divergence angles (4.02°, 9° and 13°, tested under atmospheric condition, respectively). The results are presented in Figure 10. It can be found in Figure 10 that ΔX increases with the gas temperature with approximate two patterns. In addition, ΔD changes by a smaller margin (less than 1 um). However, ΔD and ΔX satisfy a trigonometric relation. It can be also noted that the enlargement rate of the OMV volume (δ) has a small increase with the increase in gas temperature. Under the constant temperature, δ is going to increase for the larger light divergence angle, but it is still relatively smaller (<1%) compared with the OMV at atmospheric condition. Consequently, the influence of the gas temperature on the OMV can be ignored based on the above theoretical analysis.

3.2.2 Experimental method

In order to verify the reliability of theoretical analysis, an experimental dynamic model of the OMV with gas temperature variation was established based on the image analysis and the mathematical geometry relations, including the measurement of the change of light spot, the calculation of the geometric relation and the calculation of the relative variable.

By the refraction law, the changes of light spot on the inner and outer side of the glass window 2 under high-temperature condition (ΔD and $\Delta D'$) meet ΔD=$\Delta D'$. Taking the light focus position as the central axis, a triangle relation was formed by the distances between the light -

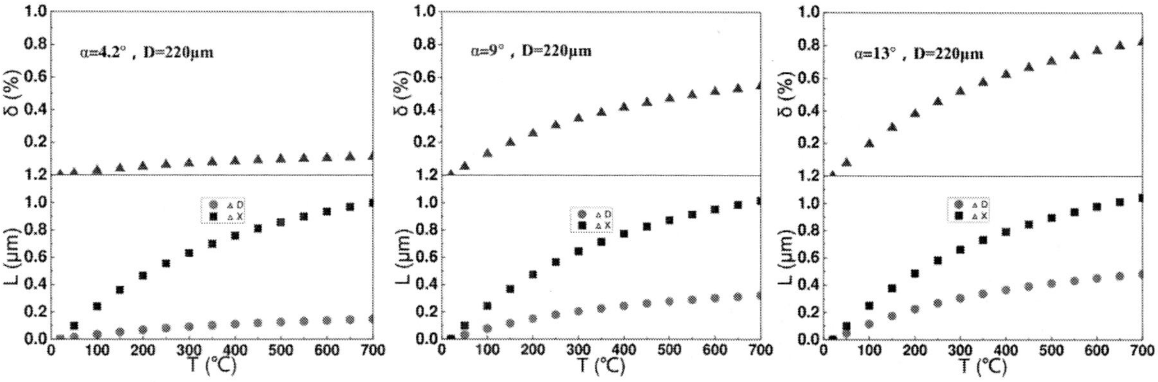

Figure 10. Results calculated based on theoretical dynamic mode with three light divergence angles (4.02°, 9° and 13°)

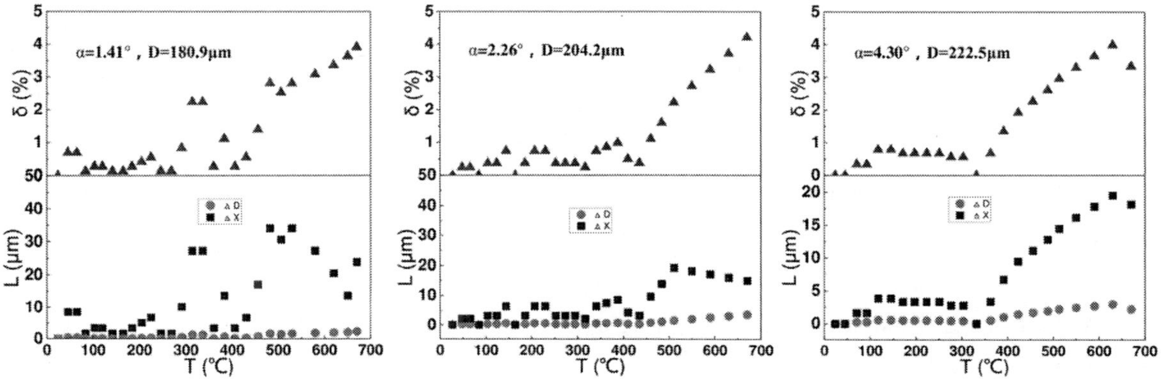

Figure 11. Results calculated based on experimental dynamic model

focus position and the inner side of the glass windows 1 and 2 (x and y), the difference of the light spot equivalent radius on the inner side of the glass windows 1 and 2 (H and h, $h = H - \frac{\Delta D}{2}$) and the equivalent radius of the OMV (H_0). Therefore, we have the following equations.

$$\frac{H - H_0}{h - H_0} = \frac{y}{x} \quad (8)$$

$$x + y = k \quad (9)$$

$$\beta = \arctan\left(\frac{h - H_0}{x}\right) = \arctan\left(\frac{H - H_0}{y}\right) \quad (10)$$

where β is the light divergence angle under high-temperature condition; k is the distance of the inner sides of the glass windows 1 and 2.

The changes of ΔX and ΔD under high-temperature condition to those under atmospheric condition are shown as follows:

$$\Delta X = \frac{y - x}{2}, \text{ or } \Delta X = y - \frac{k}{2}, \text{ or } \Delta X = \frac{k}{2} - x \quad (11)$$

$$\Delta D = 2\Delta X \tan\beta = \Delta D' \quad (12)$$

Eq. 11 to Eq. 12 are the formulas of the experimental dynamic model.

Using the above established experimental dynamic model, the ΔX, ΔD and δ of four cases in Table 2 were calculated, and the results are shown in Figure 11. It can be found that these three parameters increase obviously with the gas temperature increasing, but cannot meet the two function growth mode as the results of theoretical model. That may be caused by the reasons listed below: 1) the fluctuation of gas flow in the pipeline causes the unstable gas temperature field; 2) the beam quality analyzer is slightly influenced by other vibration sources; 3) the aerosol tube is affected by the gas flow fluctuation. But according to the experimental results, the maximum values of ΔX and ΔD are very small (less than 50 μm and 5 μm at the temperature below 700 °C, respectively) compared with the equivalent diameter of the OMV. Consequently, the ratio of the volume of the OMV is less than 5%. To sum up, the gas temperature has greatly weakened influence on the OMV, which can be ignored in actual technical development.

Table 2. Optical parameters of experimental model

number	light divergence angle /°	spot equivalent diameter /μm
1	1.41	180.9
3	1.94	285.1
4	2.26	204.2
5	4.30	222.5

3.2.3 Experiments

The performance of the optical sensor is investigated at different temperatures up to 650°C. The experimental results showed that the temperature of the optical sensor is around 50°C with the gas temperature in the range of 200-300 °C, and is going to increase about 10 °C at the gas temperature of 400-500 °C. The temperature is in the controllable range with a maximum value of 70 °C under the gas temperature of 650 °C. Therefore, the developed optical sensor can be used for the detection of particles in high-temperature conditions for a long time.

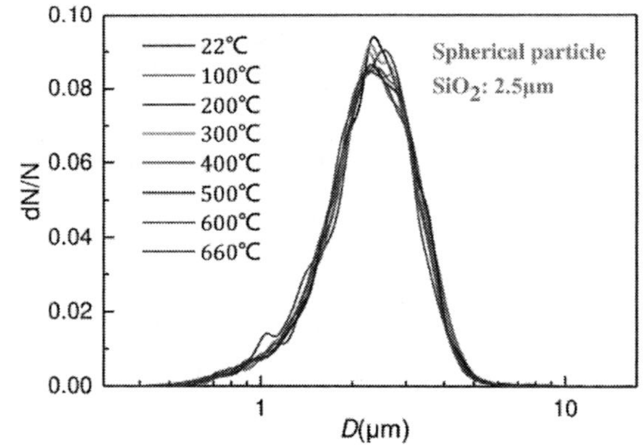

Figure 12. The results obtained under high-temperature condition

In the experiment of particle detection, the standard silica spherical particles with a strong temperature resistance ($D_{particle}$=2.5 μm) were tested at different gas temperatures. The testing time is 60 s. Figure 12 shows the test results. It can be found that all of the particle size distributions are in good agreement and have a good normal distribution, which meets the test results of single particle size, and the peak particle size is about 2.5 μm. Consequently, the optical sensor worked well for the particle detection under higher-temperature condition.

4 Conclusion

The emission of high-temperature flue gas causes serious air pollution and partial waste of available heat. In order to meet the relevant emission standards, additional purification units (high-temperature ceramic filters, etc.) are installed in the industrial process to collect the dusts in the flue gas. As the purification units are important for controlling the dust concentration, it is necessary to calculate the separation efficiency through a real-time measurement of the dust concentration in the flue gas in the upstream and downstream of the purification equipment.

In present work, a single particle measurement method and device suitable for high-temperature condition were developed. In order to complete the particles measurement at 650 °C, the beam images in optical sensing system under different temperature conditions were measured in real time by CMOS sensor. It was found that the beam images were deformed when the light divergence angle increased to more than 10°, and the deformation became more serious as the temperature increased. Depending on the principle of heat transfer, thermal expansion along the airflow direction caused by heat conduction between the duct and the gas is the main cause of this phenomenon. By changing the aerosol tube material t o one with an ultra-small linear expansion coefficient (Invar: 4J36), the optical system can be applied to the condition with temperature up to 700 °C, and in particular the light divergence angle extends to 30°.

According to the principle of gas state equation, the optical parameters (refractive index) of the gas at different temperatures change with the change in its state parameters. According to the results calculated by the formula of Ruegur's atmospheric refractive index, it was found that the refractive index of the gas decreases with the increase in temperature, which leads to the increase in the divergence angle of the beam in the aerosol tube, and shift of the position of the optical beam. Based on the geometrical optics theory and the geometric mathematical relationships, the theoretical dynamic model of the optical measuring volume with the change of the gas temperature was established. By calculating the amount of the beam focus offset and the volume of the optical measuring volume at different temperatures, it was found that the optical measuring volume increases with the temperature increasing. However, when the gas temperature was 650 °C, the increase rate of optical measuring volume was extremely small, so the influence of the temperature on the optical system can be negligible.

The performance of the whole detection system was tested by using 2.5 μm standard spherical silica particles. The satisfying measuring results were obtained and they could be applied to the actual on-site measurement at the temperature of 650 °C.

Acknowledgement

The authors thank the supports of "The National Key Research and Development Program of China (NO.2016YFB0601100)"; and would also like to thank the laboratory and staff of the Beijing Key Laboratory of Process Fluid Filtration and Separation, China University of Petroleum.

Reference

[1] J. Meyer, H. Umhauer, K. Reuter, A. Schiel, G. Kasper, M. Förster, Concentration and size measurements of fly ash particles from the clean gas side of a pressurised pulverised coal combustion test facility, Powder Technology, 180 (2008) 57-63.

[2] H. Leibold, F. Dirks, V. Rüdinger, Particulate emissions from a LLW incinerator and off-gas cleaning with a new type of ceramic candle filter, Waste Management, 9 (1989) 87-94.

[3] B.L. Perkins, Incineration facilities for treatment of radioactive wastes: a review, Incinerators, (1976).

[4] D. Fino, N. Russo, A. G. Saracco, V. Specchia, Multifunctional Filter for Treatment of the Flue Gases from Municipal Waste Incinerators, Industrial & Engineering Chemistry Research, 44 (2005) págs. 9542-9548.

[5] Scheibner, Wolters, Schumacher hot gas filter long-term operating experience in the nuon power buggenum IGCC power plant, (2002).

[6] J.M. Wang, Shell coal gasification technology and its application in China, Coal Quality Technology, (2006).

[7] Q. Xu, Z. Ji, L. Yang, Performance assessment of high temperature flue gas on-line particle size analyzer, Ciesc Journal, 63 (2012) 3506-3512.

[8] B.Y.H. LIU, V.A. MARPLE, K.T. WHITBY, N.J. BARSIC, Size distribution measurement of airborne coal dust by optical particle counters, Aihaj, 35 (1974) 443-451.

[9] R. Maus, F. Löffler, Size analysis of airborne bacteria by an optical particle counter and an impactor, Journal of Aerosol Science, 25 (1994) 141-142.

[10] B. Sachweh, H. Büttner, F. Ebert, Improvement of the resolution and the counting accuracy of an optical particle counter by fast digital signal recording, Journal of Aerosol Science, 20 (1989) 1541-1544.

[11] E.W. Max Born, Principles of Optics, Cambridge University Press, Cambridge, 1999.

[12] S. Hochrainer, Further developments of the particle counter sizer PCS-2000, Journal of Aerosol Science, 31 (2000) 771-772.

[13] B.S. Li, Z.L. Ji, L. Mei, An approach for correcting particle size distributions measured by the optical particle counter welas 2000, Aip Conference, 2010, pp. 686-689.

[14] J. Stier, M. Quinten, Simple refractive index correction for the optical particle counter PCS 2000 by Palas, Journal of Aerosol Science, 29 (1998) 223-225.

[15] H. Umhauer, Particle Size Distribution Analysis by Scattered Light Measurement Using an Optically Defined Measuring Volume, Journal of Aerosol Science, 14 (1983) 765-770.

[16] B. Sachweh, H. Umhauer, F. Ebert, H. Buttner, R. Friehmelt, In situ optical particle counter with improved coincidence error correction for number concentrations up to 107 particles cm-3 quantification and elimination of the effect of particle shape and structure, Journal of Aerosol Science, (1998) 1075-1086.

[17] D.I.J. Raasch, D.I.H. Umhauer, Errors in the Determination of Particle Size Distributions Caused by coincidences in optical particle counters, Particle & Particle Systems Characterization, 1 (1984) 53-58.

[18] H. Buettner, Measurement of the Size of Fine Nonspherical Particles with a Light-Scattering Particle Counter, Aerosol Science & Technology, 12 (1990) 413-421.

[19] D. Holve, S.A. Self, Optical particle sizing for in situ measurements Part 2, Applied Optics, 18 (1979) 1632-1645.

508245 An Experimental Investigation of Single Droplet Drying Above Boiling Point

Abstract:

An experimental method has been developed to enable dried particle structure, properties and drying rate to be predicted based on droplet drying history, particularly focusing on effects driven by boiling including bubble nucleation, growth and collapse. The method involves a single droplet drying rig which will be used to investigate drying kinetics above boiling point and across a broad range of drying conditions, drop size and composition space. The drying rig was improved as compared to previous apparatuses reported in literature, in that it is designed to measure droplet drying rates and morphological changes at high temperatures and under conditions that more closely resemble those found in industrial applications. The aim is to conduct drying experiment for droplets as small as 300μm diameter, in an air stream which temperature may be as high as 350°C.

Drying kinetics and morphological evolution will be determined, particularly focusing on the second drying stage as well as inflation/deflation periods. Initially, the constructed drying rig is checked and evaluated by conducting experiments on droplets of pure water. Afterwards, experiments will be conducted on solutions of different materials such as Hydroxypropyl Methylcellulose (HPMC), sodium sulphate and sucrose. An assessment of morphologies, transformations and drying behaviour under isothermal drying would be completed at a wide space of process condition, initial water content, initial drop size and composition space. The investigation will look to link material properties, particle size, initial moisture content and drying history to morphology. Results will be compared with single droplet drying models reported in literature.

An initial investigation has focused on mapping the morphological drying behaviour of HPMC, sodium sulphate and sucrose. It has shown that the three materials show different behaviour; sucrose showed deflation and inflation cycles, sodium sulphate droplets supersaturated and crystallized while HPMC showed inflation and rupture. Differences in drying behaviour is due to difference in material and transport properties. Therefore, the next stage of this work is focused on conducting some physical properties measurement at different water concentrations and temperature. In addition, different materials will be dried on the rig and the drying behaviour will be determined. The experimental data from this work will be used to develop, a universal morphology regime map which will link material properties, particle size, initial moisture content and drying history to morphology.

Copyright © American Institute of Chemical Engineers. All rights reserved.

508482 Characterising Powder Flow in Dynamic Processes

Abstract:

Flow inconsistency of cohesive powders is a major issue for process industries. It is often difficult to predict and control flow of cohesive powders to enable efficient and reliable process operations particularly for dynamic processes such as mixing and filling. Many quantitative and qualitative flow measurement techniques exist that are used for measuring powder flowability under quasi-static conditions. However, there are few available techniques capable of assessing powder flowability under dynamic conditions. The FT4 Powder Rheometer (Freeman Technology), which drives a rotating blade along the vertical axis through a column of powder contained within a cylindrical vessel, is one of the few available devices that is capable of measuring powder flow in a dynamic state [1]. The flow resistance of the rotating blade as it penetrates the powder bed is characterised by flow energy, however, there is lack of understanding in how to relate this measurement to process design.

This work focuses on determining the relationship between the torque and shear stress in the FT4. The effect of particle size on the external flow profile around the blade has been investigated experimentally for glass beads using a high speed camera. The material investigated shows a decrease in downward specific flow energy with increasing particle size; indicating that the increased shear resistance in the bulk for smaller particle sizes exists under dynamic conditions. Particle Image Velocimetry (PIV) analysis indicates an increase in the volume of particles displaced by the blade with increasing particle size for both upward and downward blade movement. An increase in impeller tip speed leads to a reduction in flow energy for this material, and a reduction in the volume displaced by the blade. The prevailing stresses and strains as well as the volume of particles displaced by the blade are estimated by Discrete Element Method (DEM). DEM allows for the determination of a relationship between the flow energy and/or shear stress with particle properties and blade velocity and direction. Furthermore, Positron Emission Particle Tracking (PEPT) analysis of the FT4 has been carried out to determine the radial and axial displacement of particles with respect to the bed height, and to validate the DEM results.

Copyright © American Institute of Chemical Engineers. All rights reserved.

508671 Impact ofÂ Non-Spherical Projectiles on Granular Media

Abstract:

Recently, several studies involving numerical and experimental methods have focused on the study of impact dynamics in both dry and wet granular media. An example of such instances can be an animalâ€™s foot impacting sand during its locomotion. Understanding the physics behind the interaction between granular media and animal feet can be helpful in optimizing locomotion parameters for bio-inspired robots. However, most of the studies involving impact dynamics considered spherical projectiles under different conditions, while characteristic models should involve more complex shapes. Impacting different geometries with conserved density, volume, and velocity on a granular bed may experience contrasting drag forces upon penetration. This is due to the difference in the surface areas coming in contact with the granular media. In this work, non-spherical geometries of following shapes were considered: cylinder, triangular prism, hexagonal prism and cuboid. These geometries were impacted on a loosely packed non-cohesive dry granular bed with the same impact velocities. The effect of surface area in contact with the granular media was analyzed for impact, with the help of penetration depth measurements. We also tested whether the power law for spherical projectiles is applicable for non-spherical projectile impacts.

Copyright © American Institute of Chemical Engineers. All rights reserved.

508664 Computational and Experimental Shear Cell Study with Rigid Cylindrical Particles

Abstract:

Flow properties of particles are commonly assessed through shear cells. Both the Schulze ring shear tester (SRST) and Jenike shear tester (JST) measure the interparticle stresses which are important to the design of many granular flow processes. This work uses precision-cut rigid cylinders of various aspect ratios and sizes to compare the measured stresses between both testers. It is found that the shearing results for both testers are similar, which is important for future work involving flexible biomass materials.

Previous work has sought to use shear cell results to validate various aspects of discrete element method (DEM) simulations. The SRST is recreated in a DEM simulation with rigid glued-sphere cylindrical particles to assess the contact detection method by comparing the experimental and computation results. Also, a parameterization study is performed to examine the effects of coefficient of restitution, coefficient of friction, and other particle properties on the measured steady-state stress. Overall, this serves as an efficient validation of the contact detection method which will be used to simulate complex particle shapes.

Copyright © American Institute of Chemical Engineers. All rights reserved.

508865 Defluidization Behaviour of Industrial Reactive Powders at High Temperature

Abstract:

Fluidized bed reactors are used in a wide range of industrial chemical processes, and they are particularly useful in high temperature systems thanks to their ability to provide high heat transfer rates and rapid solids mixing which lead to isothermal and more controllable conditions. Their applications spread the environmental, chemical, energy and process industries. For example, they are a key technology in the petroleum industry, catalytic chemicals synthesis processes, combustion and gasification of solid fuels (coal, wastes and biomass), metals productions and many more.

Use of fluidized beds in combustion-type operations may lead to particles sintering and bed agglomeration, which again may result in defluidization, unscheduled process downtime and additional costs. However, the details of the physical mechanisms behind such phenomena are still not entirely understood. Despite the countless researches that have been done so far, such processes still disclose unrevealed features and challenge the researchers worldwide.

The objective of this paper is to assess the effect of high temperature on the flow properties and fluidization behaviour of mixtures of carbon coke and two different titanium ores, namely synthetic and natural rutile. To this end, a simple approach for the early detection of agglomeration in the regimes of incipient fluidization -where the carbon combustion causes defluidization- will be followed. Such a monitoring approach is based on the simultaneous measurements of local temperatures and the pressure drop across the particles bed.

Copyright © American Institute of Chemical Engineers. All rights reserved.

507739 Manufacturing and Characterization of Spherical Blend PBT-PC Particles for Additive Manufacturing

Abstract:

Polymer powder-based additive manufacturing technologies often yield excellent mechanical and functional part qualities. One of these technologies is selective laser sintering (SLS), a process employing micron-sized polymer particles, which are selectively fused by a laser [1]. While hardly any boundaries regarding part design exist, there are quite some restrictions concerning the variety of commercially available SLS materials. For the time being, the most widely used polymeric material for SLS is polyamide 12 (PA12) with a market share of roughly 95% [2]. Hardly any blend materials are available; mostly , the processing of multi-material systems via SLS is achieved by simply mixing different polymer powder materials [3].

In this contribution, a new approach to blend particles is introduced. Via co-grinding and subsequent rounding, manufacturing of spherical particles consisting of semi-crystalline polybutylene terephthalate (PBT) and amorphous polycarbonate (PC) thermoplastic is possible. The polymers are jointly comminuted and mixed in a planetary ball mill and the product particles are rounded in a heated downer reactor [4,5]. The thereby achieved intermixed, but not transesterified, PBT-PC composite particles are characterized regarding size and morphology via electron microscopy. Intermixing of the polymers is investigated via staining techniques and Raman spectroscopy. Thermal properties of the blend particles are analyzed via DSC, also in regard to transesterification in SLS process-like conditions. Furthermore, flowability and size distribution of the blend particles are examined. This way, we show the high potential of our process route [4] for the production of novel polymer blend particle systems for SLS.

References:

[1] I. Gibson, D. Rosen, B. Stucker, Additive manufacturing technologies, 2. ed., Springer, New York, NY, 2015.

[2] T.T. Wohlers, Wohlers Report 2016, Wohlers Associates, Fort Collins, Col., 2016.

[3] D. Drummer, K. Wudy, F. Kühnlein, M. Drexler, Polymer Blends for Selective Laser Sintering: Material and Process Requirements, Phys. Procedia. 39 (2012) 509–517. doi:10.1016/j.phpro.2012.10.067.

[4] J. Schmidt, M. Sachs, C. Blümel, B. Winzer, F. Toni, K.-E. Wirth, W. Peukert, A novel process route for the production of spherical LBM polymer powders with small size and good flowability, Powder Technol. 261 (2014) 78–86. doi:10.1016/j.powtec.2014.04.003.

[5] J. Schmidt, M. Sachs, S. Fanselow, M. Zhao, S. Romeis, D. Drummer, K.E. Wirth, W. Peukert, Optimized polybutylene terephthalate powders for selective laser beam melting, Chem. Eng. Sci. 156 (2016) 1–10. doi:10.1016/j.ces.2016.09.009.

Copyright © American Institute of Chemical Engineers. All rights reserved.

506658 Creating Tuneable Agglomerates via 3D Printing

Abstract:

It is essential to offer a universally accepted way to systematically evaluate agglomerate breakage propensity and mechanisms. Discrete Element Method (DEM) simulation is a commonly used tool but is limited by the lack of identical, controlled agglomerates to test and validate simple models, let alone replicate the complex structure of real industrial agglomerates.

Our research presents a systematic approach to produce 3D printed agglomerates with tuneable properties. Agglomerates with different mechanical properties and structures were designed and printed by an Objet500 Connex 3D printer. The tensile and shear strength of inter-particle bonds was characterized using Instron tester. Breakage results of agglomerates with simple cubic structures and the design of agglomerates with irregular structures were presented. This 3D printing method will allow more rigorous testing of agglomerate breakage models. Discrete Element Method (DEM) has been used to simulate the agglomerate crushing using the Timoshenko Beam Bond Model (TBBM) with bond properties matching the 3D printed agglomerates. For both two agglomerate structures used in this research, the simulation and experiment showed similar breakage patterns. However, the simulation results significantly underestimated the compressive loads of spherical random structured agglomerates. In order to improve the simulations of compressive granule breakage, the influences of bond geometry definitions and non-linear behaviour of the bond contacts need to be explored.

Copyright © American Institute of Chemical Engineers. All rights reserved.

508699 Effect of Particle Size Distribution and Wettability on Penetration Behavior, Granule Formation and Granule Properties in Single Drop Granulation

Abstract:

In wet granulation, single drop granulation (where one drop forms one granule) can simplify the complexity and reveal the fundamentals of the interaction between the liquid and the powder bed. Combining drop penetration time, granule formation mechanisms and the characterization of granule properties is a novel study to achieve a comprehensive map of the single drop granulation process, which is relevant in particulate applications like pharmaceuticals, detergents, agricultural chemicals and catalysts.

In this work, single drop impact of liquid on a static powder bed was studied to investigate the granule formation mechanism, droplet penetration time, and the characterization of granule properties (morphology, internal structure and content uniformity). Water was used as the liquid binder and two pharmaceutical powders, microcrystalline cellulose (MCC) (mean particle size ~54.8 Î¼m, contact angle with water ~36Â°) and acetaminophen (APAP) (mean particle size ~9.74 Î¼m, contact angle with water ~89°), were mixed to make heterogeneous powder beds with varying compositions. The complete drop impact and penetration was recorded with a high-speed camera. The granule morphology, its internal structure and content uniformity were characterized by the prism method with image analysis, micro-CT and UV-vis spectrometry, respectively. Besides the current MCC and APAP system, MCC and APAP with other particle sizes and relative contact angles will also be studied. It is believed that the mean particle size of the powder bed is predominant in influencing the granule formation mechanism, drop penetration time, and granule properties.

Copyright © American Institute of Chemical Engineers. All rights reserved.

508017 Atomically Deposited Sintering Aids: Assessing the Effects of Alumina Particle ALD on the Sintering and Performance of SOFC Electrolytes and Dental Ceramics

Abstract:

The crystal structure of zirconium oxide (ZrO_2) can be tailored by the incorporation of aliovalent dopants, most commonly, yttrium oxide (Y_2O_3). The concentration of dopant dictates the thermodynamically stable crystal phase where 3 mol%
Y
conducting cubic phase. To achieve suitable mechanical properties and ionic conductivity, Y-TZP and YSZ ceramics must be near theoretical density, requiring high sintering temperatures. Previous researchers have successfully lowered the required Y-TZP and YSZ sintering temperatures through incorporation of aluminum oxide (Al_2O_3) particles. While this technique has proven to reduce the sintering temperature in both systems, the presence of particulate Al_2O_3 inclusions in densified parts reduces the homogeneity of the material which may have deleterious effects on densification and microstructure. In this work, we report on the use of atomic layer deposition (ALD) to conformally coat individual Y-TZP and YSZ particles with the desired amount of Al_2O_3. Constant-rate-of-heating (CRH) experiments were conducted using a horizontal push-rod dilatometer to extract the activation energy of sintering and also gain mechanistic insights into the active diffusion mechanisms as a function of ALD film thickness. Transmission electron microscopy (TEM) and scanning electron microscopy (SEM) were used to characterize the as-deposited thin Al_2O_3 films and the effects of the Al_2O_3 films on grain growth and microstructure during densification. Ionic conductivity and flexural strength tests were conducted to determine the effect of Al_2O_3 thin films on relevant material properties. Al_2O_3 incorporation by ALD is shown to have unique phenomenological effects on the densification of Y-TZP and YSZ as compared with conventional Al_2O_3 incorporation using particle-based approaches.

Copyright © American Institute of Chemical Engineers. All rights reserved.

508644 Effect of Particle Size Distribution and Operating Parameters on Conduction and Convection Heat Transfer Mechanisms in Rotary Drums

Abstract:

Granular materials undergo process steps that include transportation, drying, heating and chemical or physical conversion. In several instances, these materials need to be heated or cooled during processing, and rotary drums (kilns) are the most commonly used process equipment for this purpose. It is not always possible to conduct experiments using different materials and consider all process conditions to study the thermal mechanism due to the inherent complexity of these processes. Therefore, computational modeling and simulations have been extensively used to understand the heat transfer mechanism in rotary drums. To maintain simplicity and to overcome the challenges of simulating polydispersed particles, a common practice is to assume that the polydispersed particles with a narrow size distribution behave like monodispersed particles thermally. However, no effort has been made to validate this assumption. This work focuses on understating the effect of polydispersed particles on the conduction and convection heat transfer mechanisms in indirectly heated rotary drums. It also aims to investigate the effect of rotation speed, fill level, particle size and their interaction effects on the conduction and convection heat transfer mechanisms in rotary drums. For this, the discrete element method (DEM) (using MFIX-DEM, an open-source computational fluid dynamics solver suite) simulation technique is used to analyze the thermal behavior of spherical silica beads, which are the most common catalyst support material.

Copyright © American Institute of Chemical Engineers. All rights reserved.

507974 Selective Hydrogenation of Citral over Supported Pt Catalysts on Various Substrates

Abstract:

Highly dispersed platinum (Pt) nanoparticles (NPs) were deposited on various substrates by atomic layer deposition (ALD) for the reaction of citral hydrogenation. The substrates included multi-walled carbon nanotubes (MWCNTs), silica gel (SiO_2), commercial Î^3-Al_2O_3, and ALD-prepared porous Al_2O_3 particles (ALD-Al_2O_3). Based on TEM and CO-chemisorption analysis, the Pt NPs on all substrates were around 1.3 nm, and Pt/MWCNTs, Pt/ALD-Al_2O_3 and Pt/Î^3-Al_2O_3 showed similar Pt dispersion (75%-85%). But the Pt dispersion on SiO_2 was low (63%). In the reaction of citral hydrogenation, Pt/SiO_2 showed the highest activity due to the strong acidity of SiO_2 and the conversion of citral reached 82% after 12 hr with a selectivity of 58% of unsaturated alcohols (UA), including geraniol and nerol. Among all catalysts, Pt/MWCNTs showed the highest selectivity of UA (65%) due to its unique structure and electronic effect. In the cycling test, Pt/MWCNTs and Pt/ALD-Al_2O_3 catalysts were more stable than Pt/SiO_2 which could be attributed to the different interactions between the catalysts and the solvent.

Copyright © American Institute of Chemical Engineers. All rights reserved.

508345 Numerical Simulation of Particle Classification in a Classifier Based on Coanda Effect

Abstract:

Efficient and precise classification of particles has been an important issue in many industrial applications as well as in academia. A relatively new particle classifier based on 'Coanda effect' was developed and reported to have many favorable features (Liang 2010). They include simultaneous multiple classification, simple design with no moving parts, and easy scale up to higher production rates. However, this classifier has attracted little attention so far, and our understanding of underlying mechanism associated with classification performance needs to be improved.

The purpose of this study is to investigate the detailed fluid flow and particle transport characteristics in a Coanda classifier using CFD (Computational Fluid Dynamics). The main focus is placed on the effects of working conditions (inlet flow velocity, particle diameter) and numerical method for turbulence dispersion of particles on the classification performance. As a numerical method, three different approaches are considered: k-ε model, k-ε model combined with DRW (Discrete Random Walk) model, LES (Large Eddy Simulation). In the first approach, standard k-ε model is adopted to obtain the mean flow and then particle trajectory is calculated based on the mean flow without consideration of turbulence dispersion. However, in the second approach using DRW model, the instantaneous flow velocity is constructed under the assumptions of isotropic turbulence in order to include turbulence effect. Finally, in LES, the instantaneous flow field is dynamically and accurately obtained at each time step. Numerical simulations are carried out for two-dimensional mean flow in case of k-ε model, but for three-dimensional unsteady flow in LES. Particles are tracked using a Lagrangian approach based on one-way coupling. We consider talc particles with density of 2,860 kg/m^3 and diameter of 1 to 20 μm.

The Coanda classifier is mainly composed of three inlet channels, three outlet channels, and Coanda block. Inlet channels are divided into one main inlet and two control inlets. The main air jet containing particles is supplied with the velocity of 35 m/s through the main inlet. Two control air jets with velocities of 32 m/s (lower) and 40 m/s (upper) are also supplied through the control inlets for enhanced control.

The trajectories of moving particles obtained from the mean flow (k-ε model) are analyzed for different diameters. The main jet containing particles deflects toward the curved Coanda block due to Coanda effect as expected. Fine particles (d≤2μm) follow the flow streamlines and enter the lower outlet channel. However, coarse particles (d≥8μm) are detached from the streamlines and go into the upper outlet channel. This observation agrees well with the previous result (Liang, 2010).

Even though the mean flow and corresponding particle trajectories provide useful information, realistic particle passage do not take a form of single curve but they change in time due to flow turbulence. Therefore, particle trajectories are simulated again, turbulence dispersion being considered with DRW model. Due to turbulence dispersion, particles of the same diameter are scattered around the particle pathline obtained from the mean flow. As a result, particles of the same diameter can not be collected into a single outlet. For example, some particles (about 5%) of 1 μm enter the middle outlet even though most of them (about 95%) enter the lower outlet designed for collecting fine particles. This means that flow turbulence deteriorates the accuracy of classification.

Finally, LES simulations are performed to understand dynamic behaviors of fluid flow and particle transport. Instantaneous flow fields such as vorticity contours clearly show unsteady fluid motions inside the classifier. It is also shown that the generation of unsteady vortices leads to the scattering of particle trajectory.

A more detailed analysis including the classification efficiency will be given in the final presentation.

Acknowledgements

This research was supported by the National Research Foundation of Korea (NRF) under grant number 2017M2A8A4018482 and 2015R1D1A1A0105 9675.

References

Liang, S. (2010). *Chem. Eng. Comm.*, **197**, 1016-1032.

Copyright © American Institute of Chemical Engineers. All rights reserved.

503491 Numerical Simulation of the Secondary Air Distribution Layer in the Gas Solid Fluidized Bed Based on Flunt Software

Abstract:

Due to the continuity beneficiation in the practical production, the gangue would accumulate on the distributor forming the gangue layer, which was called Secondary Air Distribution Layer and has vital effect on the gas distribution. In this paper, the ventilation characteristics of the secondary air distribution layer was analyzed. In addition, FLUENT software was used to study the effect of the modeled layer for the fluidity characteristics (e.g., pressure drop, bed height, volatility, etc.) in the gas-solid fluidized bed with the complex layer. The results showed that the complex layer could promote the secondary distribution of the airflow in the bed, which effectively inhibited the production of large bubbles and reduced the bed pressure drop fluctuation. Thus, itâ€™s of significance to provide synergy effect on the fluidization stability in the fluidized bed. Based on the above analysis, the introduced complex structure would be convenient to achieve the coal separation effectively.

Copyright © American Institute of Chemical Engineers. All rights reserved.

507949 CFD-DEM Simulation of the Fluidization of Non-Spherical Particles in Fluidized Bed

Abstract:

In chemical and energy industries, fluidization concerning non-spherical particles (such as rod-like particle, disk-like particles, ellipsoidal particles, etc.) is often encountered, but research in this area is minimal. Yet understanding fluidization of non-spherical particles is crucial in many relevant industrial process. To investigate the fluidization of non-spherical particles, a CFD-DEM coupling algorithm is employed in this research to simulate a bubbling fluidized bed. The non-spherical particles are modeled by super-ellipsoids, multi-sphere models, and other particle models. The simulations results show that, the fluidization of non-spherical particles has significant differences from that of the spherical particles. The local void fraction and particle's projected face area perpendicular to fluid flow direction are directly related to the orientations of the non-spherical particles, which has a significant influence on the particle drag force and fluidization. Due to the significance of particle orientation, we further investigate the factors that influence the distribution of particle orientations. The results show that particle orientation depends on the fluidization time, fluidization velocity, and structure of the fluidized bed.

Copyright © American Institute of Chemical Engineers. All rights reserved.

508320 CFD and DEM Simulation of the Cold Spray Process for Surface Coating with Fine Particles

Abstract:

The coating and functionalization of technical surfaces has been become increasingly important in the last two decades for various applications, since of its manifold possibilities for process optimization, as friction reduction, corrosion resistance or improved wetting behavior. Cold spray is a thermal spraying method for surface coating. It was developed in the late 1980s in Novosibirsk [1] and has gone through continuous optimization on certain applications since then. For example showed our previous study a reduction of the friction properties on steel pins, microstructured with fine Titania particles by an adapted spraying method [2]. In the cold spray process particles are dispersed in a gas stream, which is heated up to several hundred degrees and accelerated to supersonic velocities in a Laval nozzle. The particles impact on a substrate with about 800 m/s and form a strong bonding due to the high kinetic energy which is transferred to deformation and heat in the contact zone. In contrast to other thermal spraying methods, the temperatures occurring during the process are far below the melting temperatures of the used materials, which makes the process suitable also for thermal sensitive powders. The quality of the produced coating layers depends highly on the properties of the powder and substrate as well as on the nozzle design and the process parameters as initial temperature and particle velocities. Due to the high gradients, the high occurring velocities and the small size of the particles, typically in the range of 1 to 50 µm, the contact zone of impacting particles can only be measured indirectly and with destructive methods. Therefore, still not all phenomena which lead to a strong bonding in the contact area and a high quality coating are completely understood. For the study of these mechanisms two different simulation approaches have been used:

a) Simulation of particle and gas flow with Computational Fluid Dynamics (CFD)

For a better understanding of the spray process, the nozzle, the adjacent flow areas and the substrate were computed with a CFD method. The trajectories of single particles were calculated by an Euler-Lagrange approach. For the flow the k-Ï‰ SST model was used. Local temperatures in the substrate were calculated by the law of heat conduction. The impact conditions dependent on the nozzle geometry and the particle properties. With the local temperature distribution in the substrate and in the particle, precise impact conditions could be estimated. A study has been conducted to vary the parameters of the spray process, temperature and pressure as well as the particle properties as shape factor, density and size. Moreover the nozzle geometry has been varied in the simulation to optimize the spray process. The simulation results are compared to images of coated surfaces obtained by Scanning Electron Microscopy (SEM) for validation.

b) Simulation of particle-surface impact behavior with Discrete Element Method (DEM)

The DEM is widely used for investigations of particular processes. The advantage of this method is the detailed calculation of particle-particle or particle-wall contacts to obtain the important parameters as forces and deformations. The contact force in the particle wall contact is also dependent on the surface morphology. Therefore the surface topography of two real technical surfaces, a polished and an unpolished Titanium substrate were measured by Scanning Probe Microscopy (SPM) imaging with a nanoindenter (TI Premier, Hysitron). The surface topography was discretized and implemented in a DEM model. A parameter study has been conducted to determine the distribution of forces and contact angles occurring at the particle impact. Different particle sizes and impact velocities were studied.

The cold pray process dependents on various parameters which affect each other. Since of the high gradients and the small scales occurring during the process experimental measurements are often difficult. With this numerical study different parameters could be considered uncoupled of the other parameters. The results can be used for a better process understanding and an optimization of commercial cold spray systems.

[1] Alkhimov, A. P., Papyrin, A. N., Kosarev, V. F., Nesterovich, N. I., Shuspanov, M. M.: Gas dynamic spraying method for applying a coating; US Patent, 5302414; 1994

[2] Buhl, S., Schmidt K., Sappok D., Merz, R., Godard, C., Sauer, B., Kerscher, E., Kopnarski, M., Antonyuk, S., Ripperger, S.: Surface structuring of case hardened chain pins by cold-sprayed micro particles to modify friction and wear properties; Particuology 21, 2015

Copyright © American Institute of Chemical Engineers. All rights reserved.

508568 Two-Fluid Validation of Constituive Models for the Simulation of Cylindrical Particles

Abstract:

Two-fluid simulations are a popular tool in the modeling of large processes. The constitutive models that close the governing equations are based in granular kinetic theory and typically assume that the particles are spherical. In reality, very few processes utilize spherical particles, hence there is a need for constitutive models that describe more complex shapes. Previous work used discrete element method (DEM) simulations to create a solid stress[1,2] and collisional dissipation rate[3,4] model for cylindrical particles that can replace the current spherical models. These new models have been added to MFiX and this work focuses on their validation. Cylindrical particles of various aspect ratios and solid densities were used in cratering and hopper discharge experiments. Simulations are designed to recreate the experiments in order to compare the results for crater depth and hopper discharge rate. The model effectiveness is assessed with regards to assumptions made concerning particle rotation and preferential alignment.

1. Guo Y, Wassgren C, Ketterhagen W, Hancock B, James B, Curtis J. A numerical study of granular shear flows of rod-like particles using the discrete element method. *J Fluid Mech*. 2012:1-26.

2. Berzi D, Thai-Quang N, Guo Y, Curtis J. Stresses and orientational order in shearing flows of granular liquid crystals. *Phys Rev E*. 2016;93(4):40901.

3. Buettner KE, Guo Y, Curtis JS. Using the Discrete Element Method to develop Collisional Dissipation Rate Models that Incorporate Particle Shape. *AIChE J*.

4. Berzi D, Thai-Quang N, Guo Y, Curtis J. Collisional dissipation rate in shearing flows of granular liquid crystals. *Phys Rev E*. 2017;95(5):50901.

Copyright © American Institute of Chemical Engineers. All rights reserved.

509002 The Comparsion of Coarse Grained CFD-DEM for Simulating the Dense Bubbling Fluidized Bed" Requires Immediate Action

Abstract:

CFD-DEM is an effective tool to study the hydrodynamics in dense bubbling fluidized bed. However, the expansive cost of calculating the particles motion is not easy for its application in industrial application or dense particles system. Thus several coarse grained models (CGM) were put forward to simplifying the calculation of particles and decreasing the particles motion, such as PSM, SPA, MSPA,CGEE, EMMS-DPM where a group of particles are set as a parcel or coarse particle to reduce the computation. In this study the experiment and simulation test conducted to test five coarse grained methods. The result indicated that the models increasing the gravity force, pressure gradient force and drag force linearly with the coarse grained ratio (CGR) can be effective to ensure similar macro behavior between original system and coarse graining system. The simulation data of the solid fraction distribution and bed expansion ratio reflected good agreement as experiment. But the bubbles structure became less clear as increasing the CGR. Furthermore, the power spectrum density (PSD) and Wavelet analysis of the pressure signal were used and showed that SPA and EMMS-DPM gave better agreement result in term of dominant frequency and energy ratio at mesoscale feature. Finally the computational efficiency for different models was compared and the models of EMMS-DPM, PSM and CGEE have higher computation efficiency than SPA and MSPA relatively if the collision model of spring-damping were adopted.

Copyright © American Institute of Chemical Engineers. All rights reserved.

509153 Stability Analysis of Uniform Gas Solids Flow

Abstract:

Uniform distribution of gas solids flow should be ensured to gain sufficient performance of fluidized beds. The highly chaotic entropy dissipation across solids bed will trigger non-uniform fluidization, although distributor plays an important role in obtaining homogenous gas stream. Hence, total solids pressure loading (Î¦$_T$), the pressure drop ratio between solids bed and distributor, is introduced here to present their interaction. The boundary between uniform and non-uniform fluidization can be detected by stability analysis based on Prigogineâ€™s minimum of entropy production principle, and this criterion for *N* parallel paths can be simplified into calculating the second derivative of pressure drop through one path with respect of Î¦$_T$. Furthermore, a phase diagram, illustrating the effects from operational parameter (Î¦$_T$), geometrical characteristics of distributor (C_d) and property of solids (*Ga*) on instability of uniformity, is attained to draw some useful conclusions of robust uniform fluidization.

Copyright © American Institute of Chemical Engineers. All rights reserved.

509309 Numerical Simulation of Particle Sedimentation Related to Nuclear Safety By CFD-DEM Algorithm

Abstract:

Numerical Simulation of Particle Sedimentation related to Nuclear Safety by CFD-DEM Algorithm

Byoungcheol Hwang[1], Kiyofumi Moriyama[1], and Hyun Sun Park[1*]

[1] *Division of Advanced Nuclear Engineering (DANE), POSTECH, Pohang, Korea*

**corresponding author: hejsunny@postech.ac.kr*

Abstract

The Discrete Element Method (DEM) is a technique to treat particle-particle interactions, and is used in many industries such as civil engineering, powder processing, pharmacy, and so on. Coupling of this method with Computational Fluid Dynamics (CFD) is needed in problems where the particles interacts with fluid significantly, such as bubble column reactor, mineral processing, and so on.

The simulation of important phenomena in nuclear power plant accidents that involves significant core damage and melting, i.e. so called severe accidents, is also one of the topic where the CFD-DEM is useful. After the Fukushima accident in 2011, the severe accident analysis and preparation of counter measures to mitigate the accident consequences became regulatory requirements in many countries. One of the measures for the cooling and stabilization of the molten core is the flooding of the cavity under the reactor vessel where the dropping core melt is caught in a water pool. It is expected that the molten core breaks up into particles of several millimeter sizes and accumulates on the floor, then makes a particle bed, which is hopefully coolable. In this situation, the assessment for the coolability of the particle bed is very important, and a number of experimental and analytical approached suggest that the internal/external structure of the particle bed affects strongly the coolability. The CFD-DEM is a promising candidate for an analysis method that can clarify such details of the particle bed formation.

We tested the applicability of the multiphase CFD-DEM code, an open-source developed by CFDEM®coupling in 2011, for the simulation of particle sedimentation and particle bed formation under water. Our focus in the present work is on the prediction of the particle falling characteristics and particle bed shape. Although the real condition during a severe accident includes very high temperature particles (>1000K) and boiling water (two-phase flow) in a large scale, we started with a simulation of much simple experiment cases without heat transfer or gas phase, with spherical particles only as the first step for the validation of the method. The experimental data with solid particles of various diameters (4 mm, 6 mm) and densities (3600, 6000, 7800 kg/m^3), particle release nozzle diameters (20, 30, 40 mm) were referred to. The results of numerical simulation were compared with the experimental data in terms of particle falling time, maximum particle dispersion angle, particle average falling velocity, particle bed distribution, and so on. The particle falling time was well predicted by the simulation with the maximum deviation 5%. The maximum particle dispersion angle by the simulation significantly deviated from the experimental data probably due to the influences of initial disturbances. The particle average falling velocity was overestimated. The final particle bed profile by the simulation showed agreed with the experimental observation accurately in the peripheral region, but the height of the center part was underestimated. The fluid dynamic drag model and the particle-particle contact model would need improvement for better accuracy. The present work confirmed the validity of CFD-DEM algorithm on the application for the particle sedimentation in a simple liquid pool. The particle-pool interaction including the gas phase will be the next step.

Keywords: CFD-DEM; particle-fluid interaction; particle sedimentation; nuclear safety; severe accident

Copyright © American Institute of Chemical Engineers. All rights reserved.

509440 Application of Extedned Discrete Element Method to the Melting Process of a Packed of Particles

Abstract:

Application of eXtedned Discrete Element Method to the melting process of a packed of particles

Mehdi Baniasadi, Maryam Baniasadi, Bernhard Peters, Xavier Besseron

Université du Luxembourg, Faculté des Sciences, de la Technologie et de la Communication, Luxembourg

e-mail: mehdi.baniasadi@uni.lu, maryam.baniasadi@uni.lu, bernhard.peters@uni.lu, xavier.besseron@uni.lu, web page: http:/www.xdem.de

ABSTRACT

The phenomena of melting takes place in many industrial processes, such as metal processing, environmental engineering and thermal energy storage systems. The problem of modeling of solid-liquid-gas flow including physical and chemical conversions of particles such as melting is the subject of intense research in industrial processes. These systems are difficult to model with numerical analysis due to the complex interaction between the different fluids and granular phases condering heat, mass and momentum excahnge. Basically models used to predict multiphase flows with particles are Eulerian-Eulerian and Lagrangian-Eulerian. The difference between the two models is based on different treatment of the movement of the particles. The particles are tracked individually in Lagrangian-Eulerian approach while they are considered as a continuum medium in Eulerian-Eulerian. Lagrangian-Eulerian is closer to the real physical processes rather than the Eulerian-Eulerian model for granular flows with particleâ€"particle interaction[1].

In this paper, our main focus is on the modelling of the melting process for a packed bed of particles subject to a fluid flow using a numerical technique based on Lagrangian-Eulerian so-called eXtended Discrete Elemnet Method (XDEM)[2]. In this approach, the particles are resolved as discrete phase coupled via heat, mass and momentum transfer to the surrounding fluid phase. The Discrete Element Method (DEM) coupled by heat transfer is used in order to evaluate temperature distribution, melting rate as well as velocity, oriantion and position for each particle individually. In addition the approach allows to be linked with continues numerical approaches such as Computational Fluid Dynamics (CFD) in order to solve fluid fields. The model has been validated by comparing predicted results with existing experimental data for melting of a single ice particle as well as a packed bed of ice particles in the horizontally flowing water. Fluid-particle and particle-particle intractions has also been considered. In addition, the model has the capability to be extended to the packed bed of particles with different size and properties to produce different liquid phases.

REFERENCES

[1] Dierich, F., Nikrityuk, P. A. & Ananiev, S. 2D modeling of moving particles with phase-change effect. *Chem. Eng. Sci.* **66,** 5459â€"5473 (2011).

[2] B. Peters, Thermal Conversion of Solid Fuels (WIT Press, Southampton, 2003).

Copyright © American Institute of Chemical Engineers. All rights reserved.

514248 Real-Time Interpretation of Chaotic Characteristics and Instability of Dense Gas Solid Fluidization

Abstract:

The establishment of instability monitoring system targets at analyzing the chaotic characteristics and evaluating the instability of the fluidization process for dense gas solid fluidization. Experimental pressure fluctuation and pressure drop measurements were interpreted from dense gas solid fluidization system based on chaotic time-series analysis. The pressure fluctuation signals were filtered in terms of attenuating the noise signals. The filtered pressure signals were analyzed with respect to attractor reconstruction. Based on the reconstructed attractor, the maximum Lyapunov exponent and entropy were investigated and estimated under different bed heights and air velocities. Results show that the maximum Lyapunov exponents udner different bed heights and air velocities are positive, which indicates that dense gas solid fluidization exhibits chaotic characteristics. With the increase of bed height, estimated entropy of dense gas solid fluidization increases with the increase of bed height under identical air velocity, while it varies slightly with the increase of air velocity when the air velocity alters from minimum fluidization velocity to 1.36 times of the minimum fluidization velocity.

Copyright © American Institute of Chemical Engineers. All rights reserved.

508510 Glucan Particles As Potential Carriers of Natural Flavonoids for Treatment of Idiopatic Inflammation Diseases

Abstract:

One of the novel methods of pharmaceutical formulation is the loading of the drug into biocompatible microparticles, preferably extracted from natural sources. This approach offers several potential advantages, *e.g.* increased bioavailability (the ratio of the administered drug, which actually enters the bloodstream) compared to conventional drug delivery methods, enabling the option of oral administration even for parenteral drugs. Multiple principles of the loading process are possible – drug encapsulation, adsorption onto the particle surface or precipitation inside the pores a hollow particle. The presented work focuses on precipitation of a drug from an organic solvent inside glucan particles. Glucan microparticles are ellipsoid porous shells, which are obtained from cell walls of *Saccharomyces cerevisiae* (baker´s yeast). The particle size is 2-5 micrometers. Cell walls of glucan particles are mainly composed of β-1,3-glucans. In order to prevent inflammatory reaction in organism, the yeast organelles are removed by a series of alkaline and acidic extraction steps. Subsequently, present lipopolysacharides are extracted using different organic solvents.
Lymphatic cells are mainly present in inflammation area in organism where they are responsible for progress of inflammation. Glucan particles (GPs) interact with lymphatic cell´s receptors on their surface and after that β-1,3-glucans are phagocytosed by the lymphatic cells. Therefore, GPs represent an effective oral delivery system for transporting of biological active substances. Curcumin is an anti-inflammatory flavonoid, which occurs in plants of family *Zingiberacaea.* In our project, curcumin was delivered to an inflammation in organism using GPs. Loading of curcumin into GPs was enabled via precipitation of curcumin from organic solvents. Curcumin dissolved in organic solvents was added to the GPs and through pores penetrated into cavity of glucan shell. Next, the solvent was removed and the curcumin was precipitated in glucan microparticles. GPs with the flavonoid inside were observed using confocal microscopy and SEM (scanning electron microscopy). The achieved content of curcumin was successfully analyzed by other methods, which can not be mentioned directly due to article in preparation. Furthermore, the curcumin-loaded GPs were tested *in vitro* on macrophage like cells and *in vivo* on laboratory mice in order to evaluate their effectivity. The obtained results show that these composites are promising materials for biomedical applications such as efficient treatment of inflammatory diseases (e.g. idiopatic inflammatory bowel disease, rheumatoic arthritis).

Copyright © American Institute of Chemical Engineers. All rights reserved.

508601 Use of Advanced Imaging Techniques in Tablet Disintegration Study

Abstract:

Certified testing of tablet disintegration and dissolution provides us useful information. However, there are many more that are hidden due to techniques' restrictions. Such information is e.g. mechanism of disintegration, change of surface morphology or uniformity of the substances' distribution in the tablet. Therefore, there are many techniques used to obtain the remaining information such as Magnetic Resonance Imaging (MRI), Fourier Transform Infra-Red (FTIR) or Raman spectroscopy and microscopy, Terahertz spectroscopy, Scanning Electron Microscopy (SEM) and other.

Presented work uses tablet made by direct compression of binary mixture of ibuprofen and lactose (based on previous results there was 0.2 - 0.5 w/w of ibuprofen content used) pressed at different compaction pressures (low, medium and high). The tablets were subjected to several solutions and were studied by techniques that have insight into the tablet (MRI) and ones that can see the surface (FTIR and SEM).

MRI experiments provided us with several interesting and several expectable results. As the content of ibuprofen was increasing so was increasing the dissolution time and the wetted layer was decreasing. Similar results were observed when the compaction pressure was increased. When the tablets were submerged to dissolution limiting solutions (saturated lactose solution and solution saturated with both lactose and ibuprofen) the disintegration was severely limited and some tablets began to increase their volume. However, in case of 0.2 w/w ibuprofen tablets it didn't matter whether there was dissolution discriminating solution or not, it disintegrated anyway.

FTIR microscopy showed, that not only the amount, but also the shape and size of the particles from which the tablets were compacted can be determined. These findings were further supplemented by the dry state information obtained by Scanning Electron Microscopy after different dissolution times. Further the FTIR-ATR imaging technique helped to understand the basic mechanisms of the tablet disintegration and to study the dissolution and water penetration into the model ibuprofen-lactose tablets.

Copyright © American Institute of Chemical Engineers. All rights reserved.

508632 Correlation Study between Liquid Penetration and Mechanical Properties of Pharmaceutical Tablets

Abstract:

Main phenomenon influencing treatment of patient in pharmacy is amount of active pharmaceutical ingredient dissolved in medium such as water or blood. Even though this isa key step to achieve desirable medical effect it is not the first one occurring.The first step is governed by surface properties such as wettability and roughness and the first contact point is followed by liquid penetration which is highly dependent not onlyon type of used liquid, ionstrength, liquid density but also on mechanical and structural properties of tablets such as porosity, structure strength or particle bonds.Through many analytical methods we can describe each of thesesteps individually. Aim of this study is to find proper correlation amongseveralmethods such as sessile drop method, texture analysis (TA) and magnetic resonance imaging (MRI) of tablet during dissolution.

For experiments binary mixture of tabletwas used. Tablets consist of lactose (hydrophilic) and ibuprofen (hydrophobic)with variable content (from 0.2 to 0.5 w/w ratio) and were also variable in compaction pressure used on them (from low â€“ 5 kN through 10kN as middle to 18 kN as high).Three different solutions were used according to their saturation with primary substances to prevent the tablet from dissolving. Surface properties were evaluated on contact angle and soaking rate of liquid, both measured by sessile drop method. Texture analysis was performed by recording the material resistance during dry and wet state of tablet. Then the data was processed and trends were obtained.Tablets during the MRI experiments were submerged into dissolution medium and the proceeding wetting front was measured. Later the images were processed via image analysis and trends were obtained.

Results obtained from surface experiments show that higher the ibuprofen content is the more hydrophobic surface appears to be and therefore the contact angle is increasing. Contact angle is independent on manufacturing pressure but soaking rate on the other hand depends strongly.MRI images show that this technique is not suitable for rapidly disintegration tablets, where the calculated error is within experiment conditions. On the other hand, tablet with prolonged dissolution are well measured. Results from MRI experiments also state that local saturation with substances supress dissolution of tablet. Texture analysis of tablets show that they undergo immediately structural weakening after being poured with solution independently on type of solution. Rate of this structural change mainly depends on solution's acidity. Thenthe tablet decreases its resistance in time which can be interpreted as relaxation rate. With increment of ibuprofen or saturation with primary substances this rate decrease, which means that dissolution is suppressed. Correlation between MRI and relaxation rate was done. Results shows that the correlation is mainly dependent on texture analysis measurement. On the other hand, correlation between MRI and soaking rate appears to be more tentative.

Experiment characterising surface, structural and liquid penetration properties were investigated by three different techniques. Obtained results were evaluated and put to mutual correlation. From liquid penetration point of interest there is correlation between MRI and contact angle method. Also, a correlation between MRI and texture analyser was found.

Copyright © American Institute of Chemical Engineers. All rights reserved.

508642 In Situ Drug Amorphisation By Microwave Irradiation Stabilized By Mesoporous Silica

Abstract:

One of the common issues in the pharmaceutical industry is the poor solubility of the newly developed active pharmaceutical ingredients (APIs). The solubility might be significantly enhanced by changing the form of the API from crystalline to amorphous form. The amorphous state is characterized by lower thermodynamic stability and thus better dissolution properties leading to better bioavailability. In addition, the amorphous API in a dosage form (e.g. tablets) must also remain stable and not recrystallize for at least the shelf life of the drug product (usually 2 - 3 years). Various methods are being utilized to prepare and stabilize amorphous formulations in the industry, however they usually involve difficult formulation steps or suffer from poor efficiency, low stability or can be patent obstructed.

In this work a stable amorphous formulation is prepared simply using the standard wet and dry granulation process or direct tableting. Crystalline form of API is used during the initial process. The amorphous form is then prepared *in situ* and â€œon demandâ€ either during manufacturing, or by the end user by applying microwave radiation (MW).

To stabilize the API in an amorphous form after the MW, multi-structured mesoporous silica particles were added to the formulation. The particles were prepared using an emulsion synthesis of octylamine in tetraethyl orthosilicate. Using ethanol as an inhibitor instead of the commonly used nitric acid as a catalyst reduces the silica primary nano-particles nucleation rate. The formed nuclei grow larger and then self-assemble around the octylamine droplets and form hollow micro-particles. The prepared microparticles were characterized using the Scanning electron microscopy (SEM) and the Transition electron microscopy (TEM). To characterize the porous structure, nitrogen sorption (BET) analysis was conducted. The results clearly show the particle assemblies contain macropores (cavities), mesopores and micropores and possess very high specific surface area and also very large pore volume. Together with the thermal and chemical stability of silica these properties make the particles well suitable for preserving the melted API in an amorphous form after the MW.

As a proof of concept, initially powder mixtures and later simple tablets composed of Ibuprofen (model API), silica micro-particles and a filler excipient were experimentally investigated. The *in situ* amorphization of ibuprofen was conducted using a commercial microwave oven. The heating properties of the oven, i.e. the spatial distribution of microwave heating was characterized. Various microwave power input and varying composition of the mixtures (excipient/Ibuprofen ratio) were tested with respect to the melting point of the Ibuprofen. The change from the crystalline to amorphous structure due to MW heating was analyzed by x-ray diffraction (XRD), differential scanning calorimetry (DSC) and FTIR-ATR techniques. The dissolution profiles of ibuprofen from both MW untreated and treated tablets were measured using the High Performance Liquid Chromatography (HPLC).

Based on these initial results, fully formulated tablets were prepared with various API dosage strength using a tablet press. Same compression force and same dwell time was used for all of the tablets. The final tablet composition includes excipient capable of absorption of the MW, which ensures a uniform heating in the whole volume of the tablet. Next it contains the silica microparticles, which by capillary forces absorb the API once melted and stabilize it after cooling down.

Lastly, the tablets contain standard filler, lubricant and disintegrant excipients. After the MW application the tablets change their structure due to the melted API intake into the silica micro-particles, forming visible pores in the tablet structure. This has a notable impact on the dissolution and hardness of the resulting tablets. For the final tablet characterization SEM, Texture Analysis, FTIR and Magnetic Resonance Imaging were used.

Copyright © American Institute of Chemical Engineers. All rights reserved.

508646 Manufacture of Personalised Medicines By API Printing on Porous Tablets

Abstract:

Personalised medicine offers an opportunity to produce tailor-made medicaments for patients. In order to realize this vision, it is necessary to find an alternative approach to the conventional tablet design and production. One that will replace the existing mass production of drugs offering only a few dose options of the active pharmaceutical ingredient (API), while at the same time it must meet the strict reproducibility and a defined relationship between composition and release rate of active substances requirements.

One of such approaches is to print the API on demand on a solid substrate directly when it becomes needed. Appropriate dosing of API and the possibility to even combine several APIs into a single medicament improves the variability of the dosages available for prescription and reduces the overall number of pills the patient has to take. The goal of this work is to find a simple, automated and cost-effective method of preparing personalised medicaments, which could take place directly in the pharmacy shops or at home.

The method in this work consists of a precise dosing of a dissolved API into an already prepared substrate â€“ porous placebo tablets with high sorption capacity. The tablets are prepared using direct compression in a tablet press from a simple physical mixture of a binder, filler, disintegrant and lubricant. As a filler, mesoporous silica particles were used. The high pore volume provides the free space for API sorption into the tablets and the small size of the mesopores ensure the API is deposited in an amorphous form.

The prepared particles used in the formulation were characterized using Transmission electron microscopy (TEM), Dynamic light scattering (DLS) and nitrogen sorption analysis (BET). The prepared tablets hardness and uniformity was characterized both prior and after loading. The tablets after loading were characterized using the X-ray powder diffraction and the Fourier transform infrared spectroscopy to determine the API form and deposition uniformity, followed by dissolution tests to confirm the applied dosage strength and to measure the release profile. So far, several BCS Class II APIs were successfully infused into the substrates using precise on-demand dosing. Impregnation of various doses of API were accurate. Later characterization revealed no crystalline API were present and the dissolution profiles were very satisfactory.

Copyright © American Institute of Chemical Engineers. All rights reserved.

509441 Understanding Phase Transition of Acetaminophen in the Bulk and Surface of Acetaminophen

Abstract:

Pharmaceutical solids processing often lead to the generation of new phases in the bulk and on the surface of the processed drugs. Phase transformation of the API can potentially change the physiochemical properties and influence the drug performances including solubility, dissolution and stability. To detect these changes, analytical tools such as XRD are usually employed. However, these techniques remain suitable for bulk analysis. Whereas, phase transition on the surface remains hard to detect due to the lack of literature, researches in this area and the absence of surface analytical techniques. The aim of this work is to investigate different phases (pure crystalline, partially crystalline partially amorphous and amorphous) of Acetaminophen drug by combining bulk and surface analytical techniques. For bulk analysis, FTIR and XRD have been employed. The former shows no significant differences between Acetaminophen various phases. Whereas the latter confirmed the presence of different phases indicated by a remarkable variation in the absorbance intensity the peak width. Where, high and sharp peaks obtained for the pure crystalline Acetaminophen, fewer peaks observed for semi crystalline Acetaminophen and a single broad peak observed for the amorphous Acetaminophen. For the surface analysis, Nitrogen K near-edge X-ray absorption fine structure (NEXAFS) spectra of a series of NEXAFS indicates that the precise determination of IPs and Ïƒ* shape resonance energies is sensitive enough to distinguish between the different phases of Acetaminophen. DFT calculations provides information on the molecular orbitals involved during electronic excitation occurred as a result of NEXAFS experiment.

Copyright © American Institute of Chemical Engineers. All rights reserved.

509599 Treatment of Cystinosis through Vitamin E Modified Silicone Hydrogel

Abstract:

Cystinosis is an orphan disease caused by a genetic mutation that leads to deposition of cystine crystals in many organs including cornea. Ophthalmic manifestation of the disease can be treated with hourly instillation of cysteamine eye drops. The hourly eye drop instillation is tedious to the patients leading to poor compliance and additionally, significant degradation of the drug occurs within one week of opening the bottle, which further complicates this delivery approach. Our work focuses on designing a contact lens to treat the disease with improved efficacy compared to eye drops. Our goal is to design a lens that is safe and that can deliver a daily therapeutic dose of cysteamine to the cornea while retaining drug stability. We show that cysteamine diffuses out rapidly from all lenses due to its small size. The rapid release of cysteamine and other ophthalmic drugs can be prolonged significantly by integration of vitamin E aggregates inside of the polymer matrix of silicone hydrogels. The increase in release duration increases as square of the volume fraction of the loaded vitamin E fraction which leads to high efficacy and suggests that the mechanism of attenuation in drug release is formation of nanosized barriers. The increase in duration depends on the type of the contact lens. The barriers are not effective in single-phase hydrogel lenses while the effect in the bi-phasic silicone hydrogels depends on the type of the contact lens. Vitamin E incorporation increases the release duration of both ACUVUE®OASYS® and ACUVUE® TruEye™ but the effect is more pronounced in TruEye™ likely due to the low solubility of vitamin E in the lens matrix and higher aspect ratio of the barriers. This result along with the high aspect ratio seen through SEM imaging in silicone hydrogels suggests that barriers could be forming at the interface of the silicone and hydrogel phases. The presence of vitamin E has an additional beneficial effect of reduction in the oxidation rates, likely due to a transport barrier between the oxygen diffusing through the silicone channels and drug located in the hydrogel phase. Based on this study, both Acuvue®OASYS® and ACUVUE® TruEye™ can be loaded with vitamin E to design a cysteamine eluting contact lenses for effective therapy of cystinosis. The lenses must be worn for about 4-5 hr. each day, which is less than the typical duration of daily-wear. Vitamin E loaded contact lenses could be very useful for treating many other ophthalmic diseases as well.

Copyright © American Institute of Chemical Engineers. All rights reserved.

517683 Systems Integration for Dry Granulation Based Continuous Pharmaceutical Tableting

Abstract:

Over the past decade, the pharmaceutical industry has seen significant progress in the mechanistic understanding of relevant unit operations, availability of multiple sensor technologies for inline monitoring of critical quality attributes and advances in data analytics to support manufacturing decision-making. These have enabled a systematic realization of continuous manufacturing for pharmaceutical tablets using the direct compaction route. However, the limitations and difficulty in handling of some particulate materials provide the impetus for investigating granulation routes to the continuous manufacturing of tablets.

In this work, the integration of dry granulation into a continuous tableting line is investigated. Specifically, the Process Analytical Technology (PAT) tools required for inline and at-line monitoring of the tableting line, and the semi-mechanistic models for the roller compactor and the tablet press are discussed. While the Kawakita model is used for the tablet press, the roller compaction process model is consisting of an integration of material balance and actuator dynamics from Hsu et. al (2010) with the modified Johansonâ€™s model presented by Liu & Wassgren (2016). The PAT tools, mechanistic equations, and automation system are integrated for measurement validation and data accuracy using data reconciliation in a structured ISA-95 network using DeltaV. The experimental data is presented from the operation of the continuous dry granulation line-based pilot plant at Purdue University.

Copyright © American Institute of Chemical Engineers. All rights reserved.

509248 Modeling Granular Material Mixing and Segregation Using a Multi-Scale Model

Abstract:

A multi-scale model is presented for predicting the magnitude and rate of powder mixing and segregation in blenders. The model combines diffusion and percolation correlations from small-scale discrete element method (DEM) simulations with advective flow field information from large-scale finite element method simulations. The multi-scale model predictions closely match results from both large-scale DEM-only simulations and experiments, but they are much more computationally efficient than DEM-only simulations. Therefore, in contrast to DEM, the multi-scale model is well suited for predicting blending in industrial-scale systems.

Copyright © American Institute of Chemical Engineers. All rights reserved.

Modelling Granular Media using Dynamical Density Functional Theory

T. Hurst, B. Goddard & R. Ocone

Motivation

Granular media is ubiquitous in industrial and physical problems. One of the characteristic qualities of granular media at the microscopic level is inelastic hard body collisions, which drive macroscopic behaviour. Modelling granular media is a complex and interesting problem. Solid-like phenomena make it difficult to approximate a granular system with a continuum model, while chaotic behaviour and inelastic collapse in hard particle systems introduce modelling challenges on the microscopic scale. We investigate uses of dynamical density functional theory (DDFT), augmented with data gained from microscopic models, for granular media simulation.

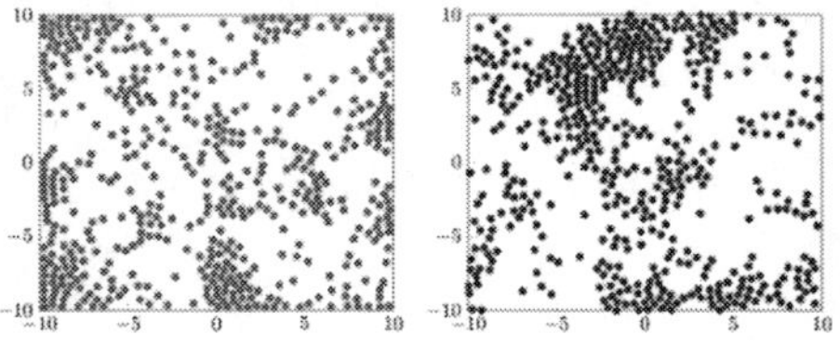

Figure 1: Inelastic systems with reflective (left) and periodic (right) boundaries

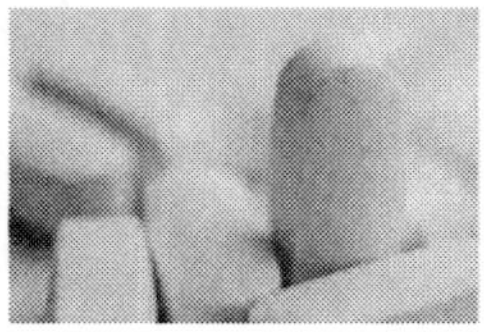

Dynamical Density Functional Theory

To derive a DDFT, we start with a set of Langevin equations for N hard particles of diameter σ in d dimensions with positions $\boldsymbol{r}^N$ and momenta $\boldsymbol{p}^N$:

$$\frac{d\boldsymbol{r}^N}{dt} = \frac{\boldsymbol{p}^N}{m}, \quad \frac{d\boldsymbol{p}^N}{dt} = -\nabla_{\boldsymbol{r}^N} V(\boldsymbol{r}^N, t) - \gamma \boldsymbol{p}^N + (m k_b T \gamma)^{1/2} \boldsymbol{G}(t), \tag{1}$$

on the hard sphere domain, $\mathcal{D}_N = \{(\boldsymbol{r}^N, \boldsymbol{p}^N) \in \mathbb{R}^{2dN} : \forall i, j, \|\boldsymbol{r}_i - \boldsymbol{r}_j\| \geq \sigma\}$, where $V(\boldsymbol{r}^N, t)$ includes external and interparticle potentials, and $\boldsymbol{G}(t)$ is an optional stochastic force term. The Kramer's equation is then a partial differential equation for its distribution function $f^{(N)}(\boldsymbol{r}^N, \boldsymbol{p}^N, t)$:

$$\frac{\partial f^{(N)}}{\partial t} + \frac{1}{m}\boldsymbol{p}^N \cdot \nabla_{\boldsymbol{r}^N} f^{(N)} + \nabla_{\boldsymbol{r}^N} V \cdot \nabla_{\boldsymbol{p}^N} f^{(N)} = \gamma \nabla_{\boldsymbol{p}^N} \cdot \boldsymbol{p}^N f^{(N)} + \gamma k_b T \nabla^2_{\boldsymbol{p}^N} f^{(N)}. \tag{2}$$

We integrate out all but one position and momenta in the Kramers equation, to find:

$$\frac{\partial}{\partial t} f^{(1)} = \mathcal{L}_1 f^{(1)} + \mathcal{L}^{2,\dots,N}_{\text{part}}(f^{(k)}) + \mathcal{L}^{(1)}_{\text{diss}} f^{(1)} + \mathcal{L}_{\text{coll}}(f^{(2)}), \tag{3}$$

where $\mathcal{L}_1$ introduces effects from external potentials, $\mathcal{L}^{(1)}_{\text{diss}}$ effects due to dissipation and friction, $\mathcal{L}^{2,\dots,N}_{\text{part}}$ effects from interparticle potentials, and

$$\mathcal{L}_{\text{coll}}(f^{(2)}) = -\sigma^{d-1} \int_{\mathbb{R}^d} d\boldsymbol{p}_1 \int_{\mathbb{R}^d} d\boldsymbol{p}_2 \int_{S^{d-1}_1} (\boldsymbol{p}_1 - \boldsymbol{p}_2) \cdot \omega f^{(2)}(\boldsymbol{r}_1, \boldsymbol{r}_1 + \sigma\omega, \boldsymbol{p}_1, \boldsymbol{p}_2, t) d\omega, \tag{4}$$

is a general collision term that arises due to the hard-sphere domain, when assuming only binary collisions. By taking the first two $\boldsymbol{p}$-moments and upon additional assumptions we arrive at a closed set of equations to model the dynamics of the system, a DDFT [1, 2]:

$$\frac{\partial}{\partial t}\rho = -\nabla \cdot (\rho \boldsymbol{v}), \tag{5}$$

$$\left(\frac{\partial \boldsymbol{v}}{\partial t} + \boldsymbol{v} \cdot \nabla \boldsymbol{v}\right) + \gamma \boldsymbol{v} = -\frac{1}{m}\nabla \frac{\delta \mathcal{F}[\rho]}{\delta \rho} - \sigma^{d-1} g_\sigma \int_{S^{d-1}_1} \rho(\boldsymbol{r} + \sigma\boldsymbol{\omega}/2)\rho(\boldsymbol{r} - \sigma\boldsymbol{\omega}/2)$$
$$\times \boldsymbol{\omega} \cdot \left\{\frac{k_B T}{m}\mathbf{1} + \boldsymbol{v}(\boldsymbol{r} + \sigma\boldsymbol{\omega}/2) \otimes \boldsymbol{v}(\boldsymbol{r} + \sigma\boldsymbol{\omega}) - \boldsymbol{v}(\boldsymbol{r} + \sigma\boldsymbol{\omega}/2) \otimes \boldsymbol{v}(\boldsymbol{r} - \sigma\boldsymbol{\omega}/2)\right\} d\omega, \tag{6}$$

where $\mathcal{F}$ is the well-studied Helmholtz free energy operator, which can include particle-particle interactions and external potentials, ρ is the density, $\boldsymbol{v}$ is the local velocity and g_σ is the value of the correlation function at σ, which must be determined. We choose to model g using statistical data from microscopic simulation.

DDFT Simulations

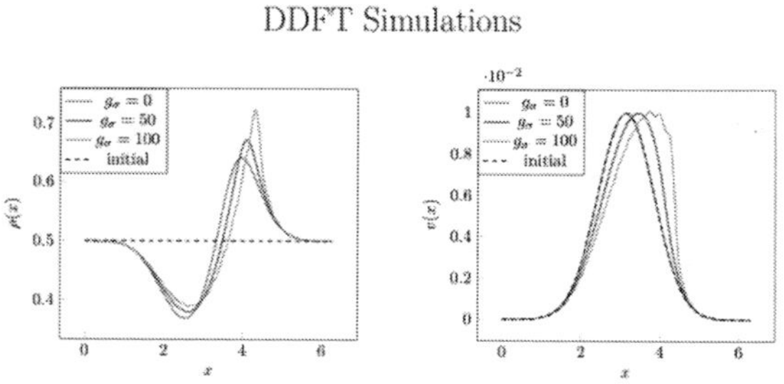

Figure 2: A simple 1D system neglecting all but the collision operator in (5-6), for different values of g. Left: density $\rho(x)$, right: local average velocity $v(x)$.

Data-Driven Modelling of g

A system of N hard spheres of diameter σ in d dimensions with positions $\boldsymbol{x}^N$ and velocities $\boldsymbol{v}^N$ obey

$$\frac{d\boldsymbol{r}^N}{dt} = \boldsymbol{p}^N, \quad \frac{d\boldsymbol{p}^N}{dt} = V(\boldsymbol{r}^N, t), \tag{7}$$

with a pairwise collision condition: if particles i and j are in contact, i.e. $\|\boldsymbol{r}_i - \boldsymbol{r}_j\| = \sigma$ and $(\boldsymbol{r}_i - \boldsymbol{r}_j) \cdot (\boldsymbol{p}_i - \boldsymbol{p}_j) < 0$, we instantaneously change the velocities via the rule:

$$\boldsymbol{p}^{out}_i = \boldsymbol{p}^{in}_i - \frac{1+\alpha}{2}\boldsymbol{\nu}_{ij} \cdot (\boldsymbol{p}^{in}_i - \boldsymbol{p}^{in}_j)\boldsymbol{\nu}_{ij}, \quad \boldsymbol{p}^{out}_j = \boldsymbol{p}^{in}_j + \frac{1+\alpha}{2}\boldsymbol{\eta}_{ij} \cdot (\boldsymbol{p}^{in}_i - \boldsymbol{p}^{in}_j)\boldsymbol{\eta}_{ij}, \tag{8}$$

where $\boldsymbol{\eta}_{ij} = \frac{\boldsymbol{r}_i - \boldsymbol{r}_j}{\|\boldsymbol{r}_i - \boldsymbol{r}_j\|}$ and $\alpha \in [0, 1]$ is the coefficient of restitution. For simple V, the Newton equations can be modelled using Event Driven Particle Dynamics (EDPD). The cell method and bounded increasing priority queues [3] reduce the computational effort to $\mathcal{O}(1)$ per collision. Using statistical methods for synthetic data (e.g. emulation [4]) we gain an understanding of the correlation function g near equilibrium, which we can include in (6-7).

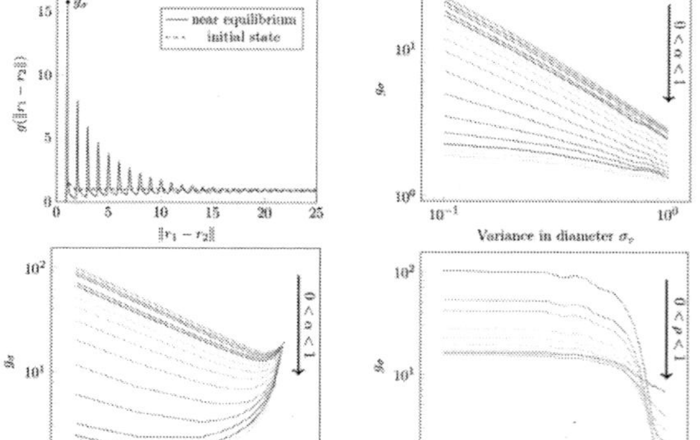

Figure 3: Top left: $g(\|r_1 - r_2\|)$ near equilibrium for a 1D system with $\rho = 0.5$ and $\alpha = 0.5$, using 10000 samples of EDPD simulations. The other three graphs show how g_σ changes for different parameters.

Future Work

We wish to include additional moments in the DDFT derivation, simulate DDFT in higher dimensions including interparticle potentials, and use EDPD in higher dimensions to calculate g.

[1] Archer, A., *J.Chem.Phys.*, **130**, 014509, (2009).
[2] Goddard, B. et. al., *Phys.Rev.Lett*, **109**, 1-5, (2009).
[3] Paul, G., *J.Comp.Phys.*, **221**, 615-625, (2011).
[4] O'Hagan, A., *Reliab.Eng.Syst.Saf.*, **91**, 1290-1300, (2006).

The support from EPSRC (grants EP/N034066/1, EP/L025159/1 and EP/L016508/01) is kindly acknowledged.

509555 A Study on Partially Wetted Particle Collisions with a Wet Wall

Abstract:

For the production of poly-ethylene, a gas-phase polymerization reaction is executed in a fluidized bed reactor. The polymerization reaction produces an excess of heat and the reactor is unable to efficiently remove the excess heat. The introduction of an Inert Cooling Agent (ICA) solves the temperature problem but also has a downside. Introducing the ICA in the reactor influences the particle collisions, leading to substantially different flow behavior. Besides the reduction of the restitution coefficient, the introduction of the liquid can result in adhesion instead of collision. This adhesion results in the formation agglomerates in the reactor and reduces the efficiency of the process. Due to the complex physics of this problem, Computational Fluid Dynamics (CFD) is a suitable method to gain insight in this phenomenon.

In order to study this complex system of partially wetted particle collisions a Direct Numerical Simulation (DNS) is used for incompressible flow. To model the coupling between the solid phases and the fluid phase, a second order accurate Immersed Boundary Method (IBM) is used (Deen et al.,2012). The interface between the fluids is tracked by a Volume Of Fluid (VOF) method (Baltussen et al., 2017). The contact-line dynamics are solved by imposing a contact angle as a boundary condition (Patel et al., 2017).

The simulated results will be compared with dry particle collisions on a wet wall (Tang et al., 2017) and with wet particle collisions on a wet wall.

References:

1. Deen, N.G., Kriebitzsch, S.H., van der Hoef, M.A. and Kuipers, J.A.M., Direct numerical simulation of flow and heat transfer in dense fluidâ€"particle systems. Chemical Engineering Science, 81 (2012): 329-344.

2. Baltussen, M. W., Segers, Q. I. E., Kuipers, J. A. M., and Deen, N. G., Cutting bubbles with a single wire. Chemical Engineering Science 157 (2017): 138-146.

3. Patel, H.V., Das, S., Kuipers, J.A.M., Padding, J.T. and Peters, E.A.J.F., A coupled Volume of Fluid and Immersed Boundary Method for simulating 3D multiphase flows with contact line dynamics in complex geometries. Chemical Engineering Science, 166 (2017): 28-41.

4. Tang, Y., Buck, B., Heinrich, S., Deen, N.G., Kuipers, J.A.M., Interface-resolved simulations of normal collisions of spheres on a wet surface. American Institute of Chemical Engineers, (2017).

Copyright © American Institute of Chemical Engineers. All rights reserved.

509716 Water Expulsion from Carbon Rods at High Humidity (Invited)

Abstract:

Porous materials including carbons, silica gels and metal organic frameworks (MOFs) are used for water management in various applications including catalysis, dehumidification and heating, ventilation and air conditioning (HVAC). Materials performance has been enhanced through chemical modifications, surface treatments or impregnation with hygroscopic salts. Regardless of the sorption behavior mechanism, all hydrophilic porous materials exhibit an isotherm in which the quantity of water adsorbed increases with an increase in relative humidity. Recently, we have produced a new class of carbon based materials that exhibit unique water adsorption phenomena. The carbon-based rods can adsorb water at low humidity and spontaneously expel about half the adsorbed water at high relative humidity. Rods synthesized under different reaction conditions (e.g., reaction temperature, and the amount of $FeCl_3$ catalyst used) have different morphology, size and exhibited different RH for onset of water expulsion. We will present the physical and spectroscopic evidence for this behavior and its potential applications.[1]

References

1 Nune, S. K. *et al.* Anomalous water expulsion from carbon-based rods at high humidity. *Nat Nanotechnol* **11**, 791-797 (2016).

Copyright © American Institute of Chemical Engineers. All rights reserved.

506317 Carbon Nanotubes to Immobilize Heavy Metals in Contaminated Soils (Invited)

Abstract:

Carbon Nanotubes to immobilize heavy metals in contaminated soils

António Alberto S. Correia[1], Martim P. S. R. Matos [2] and Maria G. Rasteiro[3]

[1] *Department of Civil Engineering and CIEPQPF – Research Centre for Chemical Processes and Forest Products, Univ. of Coimbra, Rua Sílvio Lima, 3030-788 Coimbra, Portugal*

[2] *MSc Student, Department of Civil Engineering, Univ. of Coimbra, Rua Sílvio Lima, 3030-788 Coimbra, Portugal*

[3] *Department of Chemical Engineering and CIEPQPF – Research Centre for Chemical Processes and Forest Products, Univ. of Coimbra, Rua Sílvio Lima, 3030-790 Coimbra, Portugal*

The contamination of soils with heavy metals is a growing concern in modern societies. When a soil is contaminated with heavy metals, which are inorganic, destruction of the contaminants is not an option since they are not biodegradable (Lone et al. 2008; Kroopnick 1994). To avoid the spread of contamination, soil stabilization techniques can be applied mixing materials with the soil in order to partially immobilize heavy metals. Carbon Nanotubes (CNTs) are nanomaterials known for its exceptional properties, like high surface area and adsorption capacity. Due to these unique properties, the potential use of CNTs in heavy metal contaminated water treatment has been studied, with very satisfactory results (Kabbashi et al. 2009; Abdel Salam 2013), however their application in contaminated soils is practically unexplored.

This experimental work is focused on studying the potential of using CNTs in soil remediation, especially to immobilize heavy metals. First, a large quantity of soil was collected and artificially contaminated with several heavy metals: lead (Pb^{2+}), copper (Cu^{2+}), nickel (Ni^{2+}) and zinc (Zn^{2+}), commonly present in contaminated soils. In order to avoid CNTs agglomeration, which originates the loss of their beneficial properties, an aqueous suspension of CNTs was prepared using a non-ionic or anionic surfactant combined with ultrasonic energy, to promote CNTs dispersion. Combination of two surfactants was also explored. Then, the soil, with and without the addition of CNTs, was subjected to adsorption tests to evaluate the CNTs capacity to improve heavy metals immobilization. To validate the adsorption test results, permeability tests were executed, simulating the conditions of a real case scenario.

The results obtained led to conclude that the addition of a small amount of dispersed CNTs can successfully increase the adsorption capacity of the soil and consequently improve the immobilization of heavy metals in the soil matrix (Matos 2016). The immobilization percentage varies with the type of heavy metals under study. For the soil studied, CNTs were most effective in the case of Ni^{2+} and Zn^{2+} immobilization, while in the case of Pb^{2+} and Cu^{2+} the addition of CNTs had almost no effect. Moreover, lixiviation tests conducted proved that CNTs were not released into water once they had been properly adsorbed in the soil.

Acknowledgement

The authors would like to acknowledge the financial support of Pest/C/EQB/UI0102/2013 from FCT/MCTES (PIDDAC), co-financed by the European Regional Development Fund (ERDF) through the program COMPETE (POFC).

References

Abdel Salam, M. 2013. 'Removal of heavy metal ions from aqueous solutions with multi-walled carbon nanotubes: Kinetic and thermodynamic studies', *International Journal of Environmental Science and Technology*, 10: 677-88.

Kabbashi, Nassereldeen A, Muataz A Atieh, Abdullah Al-Mamun, Mohamed ES Mirghami, MDZ Alam, and Noorahayu Yahya. 2009. 'Kinetic adsorption of application of carbon nanotubes for Pb(II) removal from aqueous solution', *Journal of Environmental Sciences*, 21: 539-44.

Kroopnick, P. M. 1994. "Vapor abatement cost analysis methodology for calculating life cycle costs for hydrocarbon vapor extracted during soil venting." In *Remediation of hazardous waste contaminated soils*, 779-90. RC Press.

Lone, Mohammad Iqbal, Zhen-li He, Peter J Stoffella, and Xiao-e Yang. 2008. "Phytoremediation of heavy metal polluted soils and water: Progresses and perspectives " In *Journal of Zhejiang University. Science. B*, 210-20. Hangzhou.

Matos, Martim. 2016. "Soil Decontamination using Nanomaterials." Msc Thesis, University of Coimbra, Coimbra, Portugal. (in Portuguese)

Copyright © American Institute of Chemical Engineers. All rights reserved.

507756 Computational Fluid Dynamic Study of Biomass Vapor-Phase Upgrading Process

Abstract:

Biomass is an important renewable energy source derived from plants like wood, food crops, agriculture or forestry residues, algae or even fumes from landfills. It is critical to find the most efficient ways to utilize the abundant biomass feedstocks. This work reports our recent efforts to model vapor-phase upgrading (VPU) reactors for biomass pyrolysis products developed within Consortium for Computational Physics and Chemistry, a research collaboration of national laboratories for the U.S. DOE Bioenergy Technologies Office. In the catalytic vapor phase upgrading process, biomass pyrolysis oils are chemically converted to more suitable products for energy or valuable chemicals. CFD modeling has been used to help understand the process and provide useful information for process scale-up and integration. The experimental tests and numerical simulations of a small-scale circulating fluidized bed system at NETL are first conducted to facilitate model calibration and validation. After that, high-fidelity numerical simulations for a pilot-scale VPU system are performed to study mixing and flow hydrodynamics under different operating conditions. Detailed information on residence times of catalyst particle and gas vapors are then extracted which can be used for the low-order process models.

Copyright © American Institute of Chemical Engineers. All rights reserved.

507882 Renewable Solvents to Replace Water in the Synthesis of Battery Materials

Abstract:

The current technologies for making metal oxide powders use either aqueous or gaseous processes. Aqueous processes involve water as solvent for metal salts to form precursor solutions. Such aqueous precursors are generally expensive for metal oxides that require high purity in which the water has to be purified. After the metal oxides are obtained, wastewater is generated and has to be treated before discharged or reused. The aqueous processes have serious environmental issues, but also are energy intensive.

Our lab has been engaged in using glycerol to replace water in the synthesis of metal oxides, especially those for use in Li-ion battery. Glycerol is a byproduct from biodiesel production. The U.S. biodiesel manufacturers generate about 1 billion pounds of glycerol yearly, which needs to be utilized for producing value added products. By using glycerol as a solvent, we found that it can replace water to prepare many metal oxide powders. The materials synthesis processes have the potential to significantly reduce manufacturing cost and alleviate the environmental issues with aqueous processes. In this talk, I will present some of our work in materials synthesis using glycerol as a solvent. In particular, I will introduce the work on making active cathode materials for the Li-ion battery.

Copyright © American Institute of Chemical Engineers. All rights reserved.

507979 Conversion of Landfill Gas to Liquid Fuels through Tri-FTS Process

Abstract:

Tri-reforming of methane is a combination of steam reforming, dry reforming and partial oxygen reforming, where steam and oxygen can tune the syngas product and decrease the carbon deposition on the catalyst. Fischer-Tropsch Synthesis (FTS) is a traditional pathway to produce liquid hydrocarbon fuels. A novel technology of combining tri-reforming and FTS (TriFTS) is proposed and utilized to convert real biogas (landfill gas) to liquid hydrocarbon fuels. In the tri-reforming section, $NiMg/Ce_{0.6}Zr_{0.4}O_2$ pellets have been developed and tested in an integrated process. The conversions of CH_4 and CO_2 were 95% and 39%, respectively. The H_2/CO molar ratio was 1.8. The $NiMg/Ce_{0.6}Zr_{0.4}O_2$ pellet catalysts contained $Ce_{0.6}Zr_{0.4}O_2$ and (Mg,Ni)O phases, and possessed the BET surface area of around 20 m^2g^{-1} through the characterization of XRD, BET, etc. In the FTS section, different eggshell catalysts (e.g. Co/SiO_2 and $Co\text{-}Fe/SiO_2$) have been developed and utilized. Using appropriate gas cleanup, the performance of the TriFTS process matched that of using the surrogate biogas. The CO conversion in FTS was 71% and the liquid hydrocarbon product matched that of low sulfur diesel. Overall, ~ 50% of the carbon in the biogas was converted to hydrocarbon products. The TriFTS process is promising for commercial scale applications. Authors affiliated with T2C-Energy, LLC disclose an interest in the technology.

Copyright © American Institute of Chemical Engineers. All rights reserved.

505718 Design of Particles and Electrodes for Electrochemical Energy Storage

Abstract:

Lithium-ion batteries have become a commercially successful battery technology; however, there is still significant room for improvement. Much progress in battery technology historically has come from advances in materials chemistry; however, other factors can have significant impact on battery performance. In particular, heterogeneity with regards to the particles that store energy in lithium-ion batteries and their organization in the electrode causes challenges with regards to maximizing energy density and keeping the electrochemical activity throughout the cell uniform.

Our group has focused its efforts on synthesizing battery active material particles with controllable morphologies. We have developed methods to tune the size, shape, and composition of precursor particles that are subsequently converted to battery active material particles. Our ultimate goal is to develop approaches to produce controlled particle morphologies using a variety of battery materials, and thus to design particles suited to a specific electrode processing or cell geometry. In this talk, the recent efforts within our group to control battery morphology and composition will be described, as well as some of our planned battery concepts.

This talk will describe advances in two areas in our lab with regards to particle and electrode design and control. In the first part, we will describe efforts to take advantage of explicit control over solution chemistry to control the composition and morphology of lithium-ion battery precursor particles. Control over precipitation of transition metals in battery precursors has been largely unreported in the open literature, and we will describe conditions where explicit control over the precipitation of transition metals is necessary to avoid the formation of phases detrimental to final battery material performance. In addition, the impact of shape modifying agents on particle composition and morphology will also be described. The second part of the talk will discuss efforts in our lab to control and characterize lithium-ion battery electrodes. For some lithium-ion battery systems, the electrode architecture and ion transport through the electrode can be the major limitation to achieving higher discharge currents. We will describe our efforts to characterize the electrode architecture, and possible routes to design particles that will help to maximize the energy density of battery electrodes.

Copyright © American Institute of Chemical Engineers. All rights reserved.

508010 Stable Cyclic Performance of $Li_{1.2}Mn_{0.54}Ni_{0.13}Co_{0.13}O_2$ Modified By Conductive CeO_2 Film

Abstract:

Surface is imperative for electrode material in lithium ion battery. Formation of solid electrolyte interface (SEI) layer, and electrochemical and chemical reactions all start from surface. As a promising technique, atomic layer deposition (ALD) has been used to provide an initially safeguarding and uniform ultrathin film with a controllable thickness. In this work, CeO_2 was deposited by ALD with different thicknesses to modify the surface of lithium rich $Li_{1.2}Mn_{0.54}Ni_{0.13}Co_{0.13}O_2$ (LRNMC) particles. The thickness was observed by transmission electron microscopy (TEM). At a condition of 1C rate and 55â„ƒ, the initial capacity of the electrode with an optimal thickness of coating exhibited 8% higher than 184 mAh/g of the uncoated LRNMC (UC) particles; there was still ~120 mAh/g retained after 400 cycles of charge/discharge, whereas only ~40 mAh/g for UC after 180 cycles of charge/discharge. A robust kinetic of electrochemical reaction was interpreted for the CeO_2 coated samples through electrochemical impedance spectroscopy (EIS). By applying AC and DC impedance spectra, conductivity of the coated samples was observed to be much higher than that of the UC at 60~140â„ƒ. CeO_2 thin coating increased conductivity of particles and blocked Mn^{2+} dissolution during charge/discharge cycling.

Copyright © American Institute of Chemical Engineers. All rights reserved.

508009 Enhanced Performance of Li-Rich Layered Cathode By CeO_2 Atomic Layer Deposition

Abstract:

Surface modification has been widely considered as an effective strategy to boost cycle-life and enhance conductivity of electrode particles for lithium ion battery. Through atomic layer deposition (ALD), a conformal ultrathin coating with controllable thickness can provide initial protection for surface of active materials. In this work, $Li_{1.2}Mn_{0.54}Ni_{0.13}Co_{0.13}O_2$ (LRNMC) was coated with different ALD cycles of CeO_2. The thickness was estimated by transmission electron microscopy (TEM). With the optimal thickness, the electrochemical performance greatly improved, with an initial capacity of 199 mAh/g, and 60.2% capacity retention after 400 cycles of charge/discharge with 1C (1C=250 mA/g) rate at 55â„ƒ, compared to uncoated LRNMC (UC) with 184 mAh/g and <22% at the same condition. These results were explained as increase of conductivity and Mn^{2+} dissolution prevention by CeO_2 coating. To support it, polarization behavior was investigated by electrochemical impedance spectroscopy (EIS). AC and DC resistance of active materials was measured by EIS and chronoamperometry (CA) to quantify the improvement of conductivity.

Copyright © American Institute of Chemical Engineers. All rights reserved.

AUTHOR INDEX

AUTHOR INDEX

AUTHOR INDEX

AUTHOR INDEX

AIChE
120 Wall Street, FL 23
New York, NY 10005-4020

ISBN 978-1-5108-6980-6

9 781510 869806

CURRAN ASSOCIATES INC.
proceedings.com